The Gardener's Guide to
Weather and Climate

THE GARDENER'S GUIDE *to* WEATHER & CLIMATE

Michael Allaby

TIMBER PRESS
PORTLAND, OREGON

Acknowledgements

I wish to thank Anna Mumford for her enthusiasm and always-cheerful support, and Michael Dempsey, my editor at Timber Press, for helping me avoid making a fool of myself. If, despite his efforts, absurd errors remain, they are my fault, not his.

Front cover photo by OGphoto
Back cover photos (from left to right) by Sermork/Shutterstock.com,
    Wikimedia/Thomas Bresson, Albert Russ/Shutterstock.com, Galyna
    Andrushko/Shutterstock.com, Denis Burdin/Shutterstock.com, Ilaszlo/
    Shutterstock.com, and Jacek Nowak/123rf

The Haseltine Building
133 S.W. Second Avenue, Suite 450
Portland, Oregon 97204-3527
timberpress.com

Printed in China
Text design by Patrick Barber
Cover design by Debbie Berne

A catalog record for this book is also available from the British Library.

Library of Congress Cataloging-in-Publication Data

Allaby, Michael.
    The gardener's guide to weather and climate / Michael Allaby.—First.
        pages cm
    Includes index.
    ISBN 978-1-60469-554-0
    1. Gardening. 2. Crops and climate. I. Title.
    SB453.A618 2015
    635--dc23   2014040762

Contents

Introduction

ANYONE WHO CULTIVATES PLANTS OUTDOORS depends on the weather, and the effects of bad weather can be dramatic. The long, cold start to 2013, for example, meant that an event held every April to demonstrate and sell mowing machinery in Dumfries, Scotland, had to be cancelled for want of grass. Bad weather destroys crops and good weather means abundant harvests. So it has been throughout history and gardeners are no less affected by the weather than farmers. There is a difference of course, in that for most gardeners a crop failure does not mean they and their families will go hungry or the creditors will come knocking at the door. Also gardeners enjoy more flexibility than is possible for farmers when deciding which plants to grow. Nevertheless, gardeners need to be able to recognize suitable conditions for sowing and planting, to know which plants to position where, and how best to help their plants survive adverse conditions. In a word, they need to understand something of how the weather works and how to predict its vagaries.

Understanding the weather means that although gardeners are affected by it they need not necessarily live at its mercy. Plants are adapted to the climates in which they grow, and adapt rapidly to climatic changes, and an informed gardener may be better equipped to make appropriate choices of species, varieties, and techniques. This book aims to help. It explains, simply and

When uneven heating makes a patch of ground hotter than its surroundings, rising air over the warm patch draws in air to replace it. Air converging toward the low pressure turns about a vertical axis and spirals upward carrying with it dust and small fragments of debris. It is then a dust devil.

succinctly, the basic principles of climatology, which is the study of climates, and meteorology, which is the study of weather, with particular reference to gardening and garden plants.

The first chapter describes the principal processes by which the action of solar radiation and the Earth's rotation and orbit produce the world's climates. It ends with an explanation of how climates are classified and a brief description of the world's main types of climate.

Climates do not remain constant. There are cave paintings showing people sailing in canoes in what is now the Sahara Desert, sharing their environment with a variety of animals including hippopotamuses, which are semi-aquatic. Thus the second chapter discusses the history of climate, leading to recent climatic changes, their likely implications, and a brief outline of past attempts to control the weather.

Climate is the average weather recorded over a period of decades. The third chapter, therefore, describes weather in more detail. It tells how differences in surface pressure combine with the planet's rotation to produce winds, and of the chain linking the compressibility of air, the formation of clouds, precipitation, and storms. It describes the different forms of precipitation, a term that includes fog, and it explains the causes of extreme events such as tornadoes and hurricanes.

This is all quite general, though, so the fourth chapter deals with microclimates, which are the local variations that matter to anyone trying to cultivate the land. It describes how friction slows the wind and produces eddies, how conditions change on hillsides, why some places are suntraps and others frost hollows. It describes where you should expect snow to form drifts, the effect shelterbelts, hedges, and fences have on the wind, and it ends by explaining why towns have a distinctly different climate from the surrounding countryside.

Wind, rain, freezing, and thawing together wear down rocks and break them into the tiny fragments that are the basis of soils, and chemical reactions among the compounds that dissolve out of the rock fragments release the nutrients that sustain soil life. The next chapter describes soils. It tells of how soils form and how they age, how they form different types, and how soils are classified. It explains how water moves through the soil and how irrigation can modify that process for the benefit of cultivated plants.

The first stage in soil formation is called weathering. The name is apt, although it is not confined to the physical actions of wind, water, and changing temperature. Nevertheless, weathering produces the substrate that supports plant life. Plants are not passive, however. They help themselves by adapting to the conditions in which they occur. The sixth chapter describes some of the ways plants have responded to climate. It describes the effects on plants of freezing and how hardiness is defined. It explains the significance of the difference between deciduous and evergreen leaves, and why spring flowers flower in spring. Deserts present special challenges for plants and the book shows how they have met them. Everyone has heard and probably seen pictures of the vast carpet of flowers that bloom within hours of a desert deluge. You will read here of how and why that happens. The chapter ends by explaining the differences between the three pathways of photosynthesis and why they are important, the difference between photosynthesis and photorespiration, and, finally, how certain plants thrive in saline habitats.

The book then returns to some of the biomes that climates produce and describes the plants most typical of some of those biomes. The short final chapter outlines a few of the ways gardeners can help their plants thrive better in good weather and survive harsh weather.

Climate

As I write this it is raining, but with some hope of the rain moving away eastward later in the day. It is mild for this time of year with snow over high ground but none at lower levels. Rain and snow, sunshine and cloud, warmth and cold, these are the phenomena that make up the weather, the conditions we experience day by day.

Arctic tundra consists of low-growing plants growing on a treeless plain. Grasses are usually present, but the predominant plants are sedges (*Carex* species), rushes (*Juncus* species), and wood rushes (*Luxula* species), with perennial herbs, dwarf shrubs, bryophytes, and lichens.

Weather is not the same thing as climate. Climate describes the average weather conditions at a particular place over many years. Typically, the weather is averaged over 30 years, a period called a climate normal, to define the climate. Someone once remarked that "climate is what you expect, weather is what you get." The two are often confused, but it is highly misleading to extrapolate the weather conditions over a short period of, say, a season or even a year or two, and imagine this represents the climate.

The world's climates result from ways in which the atmosphere and oceans redistribute the energy the planet receives unevenly from the Sun. That redistribution produces tropical storms, deserts, polar icecaps, and the sequences of depressions that track miserably across middle latitudes. But it all begins with the way Earth spins on its axis and the way it orbits its star.

How the Earth's tilted axis produces the seasons

Long, long ago—perhaps as much as four billion years ago—many scientists believe a large mass of rock struck the recently formed Earth, hitting it close to one or other of its poles. At the time, the new planet was spinning upright on

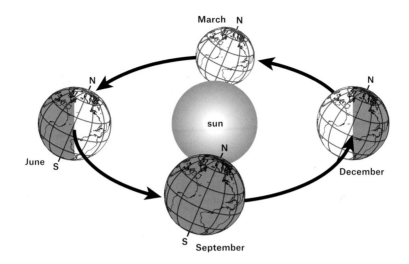

As the Earth travels around the Sun, the offset of its axis means that in June the Northern Hemisphere is tilted toward the Sun and in December it is the turn of the Southern Hemisphere. The hemisphere facing the Sun receives more intense sunlight and more hours of daylight than the hemisphere facing away from the Sun.

its axis, but the effect of the blow was to tip it to one side, so its rotational axis became tilted. If you imagine the path Earth follows in its orbit around the Sun as marking the edge of a flat disc, that disc is called the plane of the ecliptic. Without its tilt, the Earth's axis of rotation would be at right angles to the plane of the ecliptic. In fact, though, at present it is tilted at an angle of 23.5 degrees to that plane.

And that is why there are seasons. Without the tilt, if Earth travelled upright in its annual journey around its star, there would be no summer or winter, no flowers we associate with spring, no autumn colours to transform our woodlands. At Christmas we would not sing "In the bleak midwinter," for the very idea of a bleak midwinter would make no sense. For people living north or south of about latitude 40° every day would be bleak midwinter. In the British Isles and in North America as far south as about New York the days would be short and cold and the only plants able to survive would be similar to those found today to the north of the Canadian and Eurasian conifer forests, in the tundra.

The diagram above shows how the axial tilt prevents such a calamitous state of affairs by producing our seasons. As Earth travels in its solar orbit, first one hemisphere and then the other is turned toward the Sun. In June the

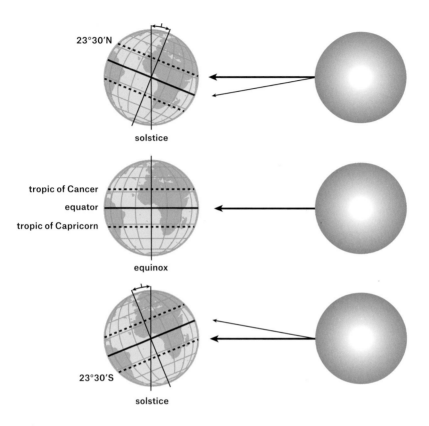

23°30'N

solstice

tropic of Cancer

equator

tropic of Capricorn

equinox

23°30'S

solstice

The angle of the Earth's axial tilt defines the location of the Tropics of Cancer and Capricorn, at 23.5° N and S.

Sun shines more brightly over the Northern Hemisphere and in December it shines more brightly over the Southern Hemisphere, giving each hemisphere its summer and winter.

Earth's tilted axis also defines the tropics and the Arctic and Antarctic Circles. As we travel the orbit that defines our year, the noonday Sun appears to wander north and south, so that in summer it rises higher in the sky than it does in winter. Consequently, summer is a time of long hours of daylight and in winter the days are short. When the number of hours of daylight, the length of time the Sun remains above the horizon, reaches a maximum, the Sun appears directly overhead at noon above the tropic in that hemisphere—the Tropic of Cancer in the Northern Hemisphere and the Tropic of Capricorn in the Southern—the names referring to constellations that are visible,

A landscape inside the Arctic Circle, where summer days are long, but the Sun is always low in the sky, so its radiation arrives at a shallow angle.

Everywhere in the tropics there are two days each year when at noon the Sun is directly overhead.

and, as the diagram opposite shows, the tropics are located at 23.5° N and S, the angle of the axial tilt. On the shortest day of the year, the Sun at noon is directly overhead above the other tropic. These two days, in midsummer and midwinter, are known as the solstices and at present they fall on 21 June and 21 December. As the Sun appears to move north and south, there are also two days in the year when at noon it is directly overhead at the equator. Those days are the equinoxes, and at present they fall on 21 March and 21 September. At the equinoxes—times of equal day and night—the Sun is above and below the horizon for the same length of time.

The contrast between the hours of daylight in summer and winter increases with increasing latitude, and in high latitudes it is extreme. The Arctic and Antarctic Circles are lines of latitude to the north and south of which, respectively, there is at least one day every year when the Sun does not rise above the horizon and one day when it does not descend below the horizon. These are the boundaries of the lands of the midnight Sun and midday darkness. This effect, too, is a result of the tilted axis. The circles are located at 66.5° N and S, that is at 90 minus 23.5°.

NORTH POLE

	0°	10°	20°	30°	40°	50°	60°	70°	80°	90°
Jan	12:04	11:21	11:01	10:14	9:22	8:18	6:23	0:00	0:00	0:00
Feb	12:04	11:29	11:13	11:06	10:25	10:04	9:07	7:12	0:00	0:00
Mar	12:04	12:02	12:00	11:34	11:32	11:29	11:25	11:17	10:31	0:00
Apr	12:04	12:13	12:22	12:32	13:08	13:26	14:19	16:04	24:00	24:00
May	12:04	12:20	13:02	13:23	14:13	15:13	17:02	22:08	24:00	24:00
June	12:04	12:25	13:12	14:02	15:00	16:13	18:29	24:00	24:00	24:00
July	12:04	12:24	13:10	13:34	14:29	15:23	17:19	24:00	24:00	24:00
Aug	12:04	12:17	12:30	13:10	13:29	14:20	15:28	18:16	24:00	24:00
Sept	12:04	12:07	12:10	12:14	12:19	12:25	13:00	13:20	15:10	24:00
Oct	12:04	11:33	11:25	11:17	11:06	10:28	10:07	9:02	5:06	0:00
Nov	12:04	11:24	11:07	10:24	10:00	9:04	7:22	3:04	0:00	0:00
Dec	12:04	11:19	10:34	10:08	9:12	8:03	5:32	0:00	0:00	0:00

SOUTH POLE

	0°	10°	20°	30°	40°	50°	60°	70°	80°	90°
Jan	12:04	12:24	13:10	13:34	14:29	15:23	17:19	24:00	24:00	24:00
Feb	12:04	12:17	12:30	13:10	13:29	14:20	15:28	18:16	24:00	24:00
Mar	12:04	12:07	12:10	12:14	12:19	12:25	13:00	13:20	15:10	24:00
Apr	12:04	11:33	11:25	11:17	11:06	10:28	10:07	9:02	5:06	0:00
May	12:04	11:24	11:07	10:24	10:00	9:04	7:22	3:04	0:00	0:00
June	12:04	11:19	10:34	10:08	9:12	8:03	5:32	0:00	0:00	0:00
July	12:04	11:21	11:01	10:14	9:22	8:18	6:23	0:00	0:00	0:00
Aug	12:04	11:29	11:13	11:06	10:25	10:04	9:07	7:12	0:00	0:00
Sept	12:04	12:02	12:00	11:34	11:32	11:29	11:25	11:17	10:31	0:00
Oct	12:04	12:13	12:22	12:32	13:08	13:26	14:19	16:04	24:00	24:00
Nov	12:04	12:20	13:02	13:23	14:13	15:13	17:02	22:08	24:00	24:00
Dec	12:04	12:25	13:12	14:02	15:00	16:13	18:29	24:00	24:00	24:00

The hours of daylight vary with latitude and season. At the equator (latitude 0°) the Sun is above the horizon for 12 hours and 4 minutes every day of the year. At the North and South Poles (latitude 90°), it does not rise above the horizon in midwinter and is above the horizon for 24 hours a day in midsummer.

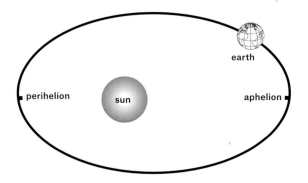

Earth's orbit is eccentric, so its distance from the Sun varies through the year. At present it is at its closest (perihelion) on about 3 January, and most distant (aphelion) on about 4 July.

It follows, therefore, that the hours of daylight change through the year, increasing from a minimum at the winter solstice to equal day and night at the equinoxes, and reaching a maximum at the summer solstice. It also follows that the hours of daylight vary with latitude. At the equator (latitude 0°) the Sun remains above the horizon for about 12 hours every day of the year, but at the North and South Poles (latitude 90°) it does not rise above the horizon at all in midwinter and it does not sink below the horizon at all in midsummer, although even then it remains low in the sky. The table opposite shows the mean hours of daylight for each month for every 10 degrees of latitude.

Just to complicate matters a little further, the Earth's orbit is elliptical. Astronomers describe such an orbit as eccentric and the extent to which the orbit deviates from a circle is its eccentricity, which can be measured. An ellipse has a centre (C) and two foci (F_1 and F_2). The distance between F_1 and C is the linear eccentricity, le, and the distance between C and the point on the circumference farthest from F_1 is a. The eccentricity (e) is equal to le/a. At present the Earth's eccentricity is 0.017.

This is important because the Sun lies at point C and Earth orbits about one of the foci of the ellipse—the orbit is eccentric because it is about a point that is not at the centre of the ellipse. Sometimes, therefore, Earth is closer to the Sun and sometimes it is farther away. It is closest at perihelion and farthest at aphelion. At present, perihelion is on about 4 January and aphelion on about 4 July. The diagram above illustrates the effect.

At aphelion the Earth receives about 7% less solar radiation than it does at perihelion. You might expect, therefore, that Northern Hemisphere summers are cooler than they might be. So they would be, were it not that another factor overwhelms the effect. There is much more land in the Northern than in the Southern Hemisphere, and land warms and cools faster than water, so although Earth receives less solar radiation in the northern summer, the Northern Hemisphere absorbs it faster. The Southern Hemisphere, in contrast, has much more ocean, which absorbs warmth slowly, so it warms less in its summer despite receiving more sunshine. The result is that the distribution of land and ocean cancel out almost all of the 7% difference between aphelion and perihelion.

You may have noticed that I've qualified all of the dates for the solstices, equinoxes, aphelion, and perihelion by saying "at present." That is because all of them change over very long periods due to the fact that the angle of the Earth's axial tilt also changes. The tilt means that the Sun's gravitational attraction pulls slightly more strongly on the pole inclined toward the Sun and slightly more weakly on the opposite pole. This has the effect of making the axis wobble. You can produce the same effect by gently nudging the top of a spinning toy gyroscope. The Earth's tilt varies from a minimum of 22.1 degrees to a maximum of 24.5 degrees over a period of 22,000 to 26,000 years, causing a line produced from the axis to describe a cone. This "wobble" is known as precession and it affect the dates of aphelion and perihelion, and consequently the position of the Earth in its orbit at the dates of the equinoxes and solstices. This gradual change is known as the precession of the equinoxes. Obviously, a change in the angle of tilt will affect the world's climates. Other changes in Earth's movement also affect the global climate. Together they trigger ice ages and the interglacials that punctuate them.

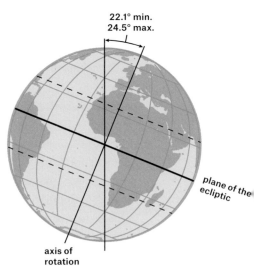

The Earth's axial tilt varies from 22.1 to 24.5 degrees over a period of 22,000 to 26,000 years.

Why summer is warm and winter is cold

In summer, the farther you are from the equator, the longer are the days. In the north of Scotland you can read a newspaper outdoors at 11 PM provided the sky is not overcast. Down south in the tropics the days are about seven hours shorter. Why then, since it is sunshine that warms us, is Scotland not much warmer in summer than, say, Venezuela?

The diagram opposite provides a clue. It shows that radiation from the Sun travels in parallel waves, but that in high latitudes, where the Sun is always low

The higher the Sun is in the sky, the more intense the radiation reaching the surface. In high latitudes the sunshine is spread more diffusely over a larger area.

in the sky, its rays strike the surface more obliquely than they do in low latitudes, where the Sun is higher. The rays from the low Sun illuminate a larger area than those from the high Sun, but they do so less intensely. The energy is spread more thinly, as it were. In summer northern Scotland receives more hours of sunshine than Caracas, but the sunshine is far less intense. Weak sunshine is better than no sunshine at all, of course, and even in Scotland summers are warmer than winters and days are warmer than nights.

While the Sun is shining on the surface—any surface—the surface will absorb some of the radiation. It will not absorb all the radiation because some is reflected, the proportion depending on the colour of the surface. The reflectivity of a surface is known as its albedo and it is reported as a percentage (for example, 10%) or decimal fraction (0.10). The table shown here gives the albedo of a variety of surfaces.

Its low albedo explains why you often come across sheep lying on country roads. The road reflects less of the radiation falling on it than does the adjacent field or woodland, so the road feels warmer. The high albedo of freshly fallen snow also explains why it takes a prolonged period of warm sunshine to melt lying snow.

Snow can be a boon to small animals such as mice and voles, however, because it insulates the ground beneath, preventing it from losing heat. It also allows them to scuttle about through tunnels, hidden from predators.

Surface	Albedo
fresh snow	0.75–0.95
old snow	0.40–0.70
cumuliform cloud	0.70–0.90
stratiform cloud	0.59–0.84
cirrostratus	0.44–0.50
sea ice	0.30–0.40
dry sand	0.35–0.45
wet sand	0.20–0.30
desert	0.25–0.30
meadow	0.10–0.20
field crops	0.15–0.25
deciduous forest	0.10–0.20
coniferous forest	0.05–0.15
concrete	0.17–0.27
black road	0.05–0.10

It may sound paradoxical to suggest that snow insulates the ground beneath it. How can a blanket of frozen water prevent heat loss? The answer is that although snow reflects incident sunlight it also traps radiation from below. We all know that temperatures will plummet on a clear winter night but that overcast skies mean warmer nights. The blanket of snow acts in the same way as do clouds.

Sunlight is electromagnetic radiation and the Sun emits radiation across a spectrum that ranges from gamma and X-rays at the high-energy, short-wavelength end to radio waves at the low-energy, long-wavelength end. Gamma and X-ray radiation do not penetrate the atmosphere. Visible light, the part by which we see, is about in the middle, with ultraviolet just beyond it on the high-energy side and infrared on the low-energy side. Regardless of the wavelength, when radiation strikes the surface, molecules absorb it and the energy they acquire makes them vibrate more vigorously. In that way the radiation is converted to heat, warming the surface.

When any object is warmer than its surroundings it emits radiation until its temperature has fallen to that of its surroundings. In other words, place a warm object somewhere cool and its temperature will fall, even in a vacuum where it cannot lose heat by convection. In 1879 the Austrian physicist Josef Stefan discovered the relationship between the temperature of an object and the amount of energy it radiates, and in 1884 Stefan's student Ludwig Boltzmann qualified that relationship. It is now known as the Stefan—Boltzmann law, expressed as $E = \sigma T^4$ in joules per square metre per second, where E is the amount of energy, σ is the Stefan—Boltzmann constant (5.6703×10^{-8} watts per square metre per kelvin), and T is the temperature of the object.

During the hours of daylight, even when the sky is overcast, the Earth's surface absorbs solar radiation and its temperature is higher than that of its surroundings—empty space. Consequently, even as it absorbs radiation it also emits radiation. The wavelength at which a body radiates is inversely proportional to its temperature, so the very hot Sun radiates most strongly at short wavelengths and Earth, which is fairly cool, radiates at long wavelengths, in the infrared part of the spectrum. The surface absorbs radiation faster than it reradiates it, however, so its temperature rises. But when the Sun sinks low, the amount of incoming, shortwave radiation decreases until, during the hours of darkness, it ceases altogether. The surface no longer absorbs radiation but it continues to emit radiation through the night, so its temperature falls. That is why the ground warms by day and cools by night, and it is contact with the ground that determines the temperature of the low-level air. An hour or so before dawn the sky begins to grow lighter and the first sunlight of a new day reaches the surface. The surface temperature ceases to fall and as the Sun rises above the horizon it starts to rise.

By extension, the balance of incoming and outgoing radiation is also part of the reason why summers are warm and winters cold. For as long as the hours of daylight exceed those of darkness, the Earth's surface absorbs more radiation than it loses, so its temperature rises day after day. Once past the autumn equinox, as the days grow shorter and the nights longer, the surface loses more radiation than it gains, its temperature falls, and before long people remark that winter is approaching.

How circulating air and water transfer heat

The Moon resembles the Earth in several ways. It is much smaller, of course, but like Earth it is made of rock and it is about the same distance from the Sun. The Moon spins at a different rate, one lunar day being equal to 27 Earth days, but all the same you might expect that it would have a climate not too different from those on Earth. But, of course, that is entirely not the case. During the lunar day the temperature can reach 107°C (225°F) and at nighttime it can fall to a distinctly unearthly –153°C (–243°F).

Even allowing for such extremes of temperature, the Moon does not have a climate. There is no weather, no wind or rain, not even a dust storm, and without weather there can be no climate. The Moon cannot have weather for it possesses no atmosphere to deliver moving air and no oceans to moisten that air. It is Earth's atmosphere and oceans that generate climates, and in doing so they moderate the temperatures.

Just as the difference between daytime and nighttime and summer and winter temperatures is due to the balance between the energy the surface absorbs and radiates, so the entire Earth experiences an energy imbalance that movements of air and water correct. From the poles to about latitude 45°, the Earth's surface loses more energy as longwave radiation than it absorbs from sunshine, and between 45° N and 45° S it absorbs more energy than it loses. But the scientists who worked out how air circulates on a global scale knew nothing of the energy balance. What concerned them were the winds.

In the seventeenth century, as well as studying the stars astronomers were expected to study weather and tides, and one of the puzzles they grappled with concerned the trade winds, important to mariners because they were so reliable, most of the time blowing from the northeast to the north of the equator and from the southeast in the Southern Hemisphere. *Trade* used to

Warm air rising on the low-latitude side of the Hadley cells is very moist and as it rises its moisture condenses, producing heavy rainstorm.

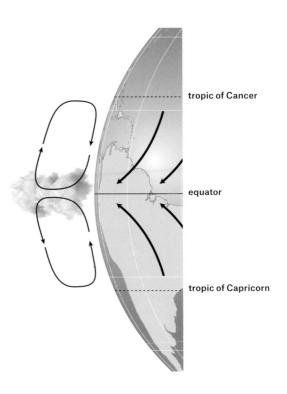

In a Hadley cell, warm, moist air rises over equatorial regions. As it rises, the air cools and its moisture condenses to form towering clouds. The air, now cool, moves away from the equator and subsides in the subtropics, warming by compression as it sinks, and flows toward the equator.

tropic of Cancer

equator

tropic of Capricorn

mean a regular track, and hence the name. In 1686 the English astronomer Edmund Halley proposed that warm air rises over the equator and the trade winds are the cool air drawn in to replace it. This was partly correct, but he had failed to account for the fact that the winds approach the equator from the east. In 1735 the English meteorologist George Hadley believed he had the answer. He suggested that warm air rises at the equator, but that the cool air flowing in to replace it is deflected by the Earth's rotation. He was very nearly right, but we know now that the convection cells he described and that bear his name do not extend all the way to the poles, and there are several Hadley cells in each hemisphere.

Hadley cells form the first part of the circulation by which the atmosphere transports heat away from the equator. As the diagram above illustrates, air, warmed strongly by contact with the surface, rises to about 16 kilometres over equatorial regions. Oceans cover most of those regions, so the rising air is very moist. The air cools as it rises and its moisture condenses to form towering clouds that deliver abundant rain. The much drier air moves away from the equator and subsides in the subtropics, around latitude 30°, warming by compression as it sinks.

Over the poles, air chilled by contact with the frozen surface contracts and draws air downward. The air flows away from the poles, rises at about latitude

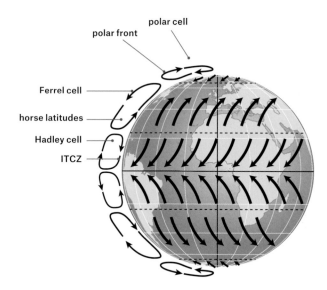

polar cell

polar front

Ferrel cell

horse latitudes

Hadley cell

ITCZ

The three-cell model is a much-simplified description of the way atmospheric movements transport warmth away from the equator. Hadley cells carry warm air to the subtropics and cold air sinks to the surface in the polar cells. The Hadley and polar cells drive the midlatitude Ferrel cells. The arrows indicate the directions of the low-level prevailing winds. There are fronts at the boundaries of the cells, the polar front being the most important, and the intertropical convergence zone (ITCZ) is where air from the Northern and Southern Hemispheres converges and rises, producing regions of light winds, known as the horse latitudes.

60°, and returns to the poles. This circulation forms a second set of cells, the polar cells.

There the explanation remained until 1856, when the American climatologist William Ferrel published a mathematical model of the atmospheric circulation that added a third set of midlatitude cells, now called the Ferrel cells. These are different, because unlike the Hadley and polar cells they are not driven directly by differences in temperature. Instead, they are driven indirectly, by rising and subsiding air in the cells to either side. Together, these sets of cells comprise a simple description, the three-cell model, of the way air circulates.

The three-cell model should not be taken literally. The real atmosphere is a great deal more complicated. Nevertheless, it gives a useful sketch of the way the atmosphere transports heat, and its principal features are real enough. The boundaries between each of the sets of cells are fronts that have a profound influence on surface weather. The intertropical convergence zone (ITCZ), where the trade winds converge forcing air to rise, moves north and south of the equator with the seasons, bringing summer rains, and subsiding air on the poleward sides of the Hadley cells produces regions of high pressure, the subtropical highs, where the winds are light and variable. These came to be known as the horse latitudes because sailing ships becalmed there often ran short of drinking water and horses carried as cargo sometimes died of thirst and were thrown overboard.

As air flows at a low level toward or away from the equator it is deflected, just as Hadley suggested, to the right in the Northern Hemisphere and to the left in the Southern Hemisphere. This deflection produces belts of prevailing winds. Those of midlatitudes are westerlies and easterlies blow in high latitudes and in the tropics—the trade winds. Winds exert a pressure on the surface and the global easterlies and westerlies balance. If they did not, the Earth's rotation would accelerate, making the days shorter, or decelerate, making them longer. And the winds also help in the transport of heat, by driving the warm and cool ocean currents.

Ocean currents and gyres

Poolewe is a village of 200 people on the west coast of Scotland, about 120 kilometres northwest of Inverness and at latitude 57.77° N. It is a popular tourist attraction, partly because the village is very pretty and the coastal scenery splendid, but mainly because of Inverewe Garden, which is filled with subtropical plants. These exotic plants thrive, thousands of kilometres north of the subtropics, because of the mild climate. The average January minimum temperature is 2°C (36°F), so frost and snow are rare. In truth, the Poolewe climate is only slightly milder than that of Inverness, latitude 57° N on the eastern side of the country, where the average minimum temperature in January is 1°C (34°F), but that one degree makes all the difference. Why is Poolewe warmer than Inverness? Ask anyone and they'll tell you it's the Gulf Stream.

They are almost correct. The Gulf Stream, part of a system of ocean currents carrying warm water northward from the Gulf of Mexico, crosses the Atlantic in the latitude of Portugal, then turns south, becoming the cool Canary Current as it passes Africa and then the North Equatorial Current, flowing westward to the Gulf of Mexico and completing the circuit. But as it turns, so the current divides, and one branch heads northeast as the North Atlantic Current, also called the North Atlantic Drift. This warm current flows close to the west coast of Ireland and clips the Scottish coast near Poolewe before heading toward northern Norway as the Norway Current. The current does not affect the climate by bathing the land in warm water, but by warming air approaching from the west.

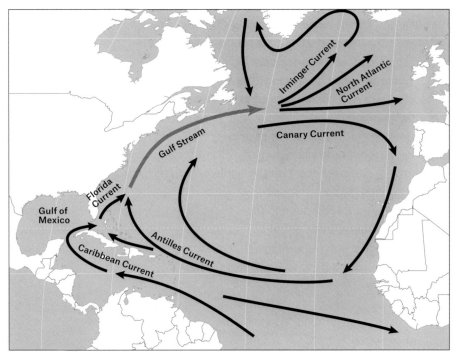

The Gulf Stream is part of a system of ocean currents that circulate in the North Atlantic. As well as the Gulf Stream the system comprises the Antilles, Florida, and Canary Currents, with the Irminger and North Atlantic Currents as branches.

The above map shows the system of currents of which the Gulf Stream forms part. The North Equatorial Current turns to the right as it approaches North America, and the Caribbean and Antilles Currents join it to form the Florida Current. Where the North Atlantic Current separates as one branch, another branch flows toward Iceland and then Greenland as the Irminger Current, and the main current turns south as the Canary Current.

The Gulf Stream is like a river of warm water flowing through the cooler water of the ocean. Its temperature is fairly constant, at 18–20°C (64–68°F) and it moves at an average 6.4 km/h, but in places up to 9 km/h, transporting 113 million cubic metres of water a second.

Clearly, the Gulf Stream system carries so much water that it is a highly effective distributor of warmth northward from the equator. And it is not alone, of course. Similar currents flow across all of the oceans. Those in the South Atlantic, for instance, comprise the South Equatorial, Brazil, South Atlantic, and Benguela Currents, and in the South Pacific there are the South

Sargassum is a genus of floating seaweed that ocean currents concentrate in the Sargasso Sea. The seaweed often drifts ashore, as here, along the coast of the Gulf of Mexico and the Florida Keys.

Equatorial, East Australia, South Pacific, and Peru Currents. Other currents, such as the Labrador off northeast Canada and the Oyashio off northeast Asia, bring cold water to join the main systems. And, as the map (on page 28) of ocean currents shows, all of the principal currents flow in circles, called gyres.

It is the prevailing winds that drive ocean currents, pushing the surface water forward. As the water moves, it is deflected to the right in the Northern Hemisphere and to the left in the Southern Hemisphere. In the case of the North and South Equatorial Currents, which are driven by the trade winds and flow in a westerly direction on either side of the equator, the deflection

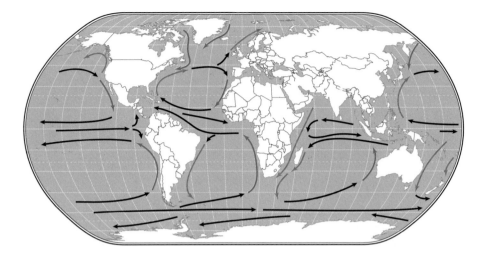

The prevailing winds drive surface ocean currents that transport warm water away from the equator and cool water toward the equator. Deflected by the Coriolis effect, in each ocean the currents follow an approximately circular path, called a gyre.

causes them to turn away from the equator. As they do so, the deflection becomes stronger, turning them further until they enter the region of the midlatitude prevailing westerlies, which drive them in an easterly direction. Again, they are subject to the deflection, now turning them back toward the equator where they rejoin the equatorial currents. That is why the ocean currents describe circles.

A circular system that transports warm water away from the equator must also transport cool water toward the equator. Consequently, the ocean gyres have two sides, one warm and one cool. An ocean current that flows parallel to the coast of a continent and fairly close to it is known as a boundary current, so in each ocean there are western and eastern boundary currents. Western boundary currents, such as the Gulf Stream in the North Atlantic and the Kuroshio in the North Pacific, are narrow and fast moving—the Kuroshio flows at up to 11 km/h—and carry warm water. Eastern boundary currents, such as the Canary in the North Atlantic and California in the North Pacific, are wider, slower, and carry cool water.

Contact with the warm or cool water surface affects air that crosses over a boundary current. The warm North Atlantic Current, part of a western boundary current despite being on the eastern side of the ocean, makes it possible for subtropical plants to grow at Poolewe in northern Scotland. Fog often forms in air that crosses an eastern boundary current, as its temperature falls and its water vapour condenses. The fog may then roll inland. This type of fog is known as advection fog and it is well known in San Francisco, caused when warm air from the ocean crosses the California Current.

Teleconnections

Thus far, the behaviour of the atmosphere seems fairly straightforward. Warm air rises in one latitude, cool air subsides in another, and three sets of convection cells deliver rain to some regions and deny it to others. Prevailing winds, generated by the vertical circulation of the three cells, drive ocean currents that circle the ocean basins, assisting in the distribution of heat. If only it were so simple. Then anyone could forecast the weather. The seventeenth of July? Sunny and warm in Manchester, raining in Miami. You could look it up in an almanac.

Unfortunately, the real atmosphere is more complicated. Its behaviour is chaotic, which means that a very small event in one place can produce effects thousands of kilometres away, over a different ocean or continent. The most famous example of this occurred in 1961, when the American mathematician and meteorologist Edward Lorenz was running a computer model that made weather predictions. This was in the very early days of computer modelling and Lorenz was experimenting to see how the technique might be applied to weather forecasting. When preparing to repeat a run of his model, Lorenz cut a corner by entering a particular value as 0.506 rather than the full 0.506127, then left the room while the computer produced its result. When he returned he found it had produced an entirely different weather pattern, a change he traced back to that abbreviated data entry. When describing his work at a meeting in 1972 his talk bore the title "Does the flap of a butterfly's wings in Brazil set off a tornado in Texas?" Since then, the result of such extreme sensitivity to starting conditions has been known as the butterfly effect and the mathematics describing it became chaos theory. It is what makes it impossible to predict the weather for more than a week or two ahead. Long-range forecasting is not merely difficult, but impossible in principle because the variations that cause apparently identical patterns to diverge are too small to be detected.

Some of these long-range effects are more regular, however. There are certain variations in the distribution of air pressure and circulation that span vast geographical areas such as an entire ocean or continent and that persist for weeks to months, or even to several consecutive years. Such linkages are called teleconnections and there are many of them.

There is one pattern, for example, that involves the distribution of pressure between the Kamchatka Peninsula and part of southeast Asia and the subtropical western North Pacific, so that when one region has high pressure, pressure is low at the other. The effect is that when winter and spring temperatures are above average in the subtropical western North Pacific they are below average all year in eastern Siberia, while places farther north have wetter weather than usual all year, but the weather is drier than usual over the central North Pacific in winter and spring.

The Pacific–North American teleconnection pattern is linked to lower-than-average pressure over Hawaii and the intermountain region of North America and higher-than-average pressure over the Aleutian Islands and the southeastern United States. When the pressure differences are pronounced, western North America experiences warmer weather than usual, and it is cooler than usual in the south-central and southeastern United States. The weather is wetter than usual in Alaska and the northwestern United States and drier than usual in the Midwest.

From 1950 to 1976 temperatures all year were below the long-term average across Europe and above average in the southern United States between January and May and in the north-central United States from July to October. The weather was drier than usual in northern Europe and wetter than usual in southern Europe. Then the pattern reversed from 1977 to 2004. This was due to another teleconnection pattern and yet another of these patterns links warm conditions in eastern Asia with cool conditions in western Russia and northeast Africa.

Teleconnections involve weather patterns over large distances that are caused by external variations. That is to say, warm weather in one place may be linked to cool weather somewhere else, but the warm weather did not cause the cool weather or vice versa. Both were caused by something else. If it is possible to identify weather patterns that are linked over large distances in this way it may be possible to predict well in advance the disasters they can trigger, especially floods and outbreaks of mosquito-borne diseases. With that end in view, scientists are working to relate historical records of teleconnection patterns to records of major floods.

El Niño

There are many cyclical changes in climate that affect the weather far from the regions where they occur, but El Niño is by far the most famous, perhaps because of its intriguing name (Spanish for male child), which refers to the

Christ child because an El Niño event usually reaches its climax in December. It is linked to a change in the distribution of surface air pressure over the South Pacific, measured at Darwin, Australia, and Tahiti. Ordinarily, pressure is low in the west (Darwin) and high in the east (Tahiti). Surface winds blow from east to west—from high to low pressure—and high-level winds blow in the opposite direction. At intervals of two to seven years, however, the pattern changes. Pressure is then high in the west and low in the east, and the winds weaken or change direction. This periodic change is called a southern oscillation.

Ordinarily, water off the west coast of tropical South America is cool. The western boundary Peru Current, carrying water from the Southern Ocean, flows parallel to the coast with many upwellings that bring cold water to the surface. The Sun beats down strongly, of course, but the water it warms is swept westward by the South Equatorial Current, driven by the trade winds, and accumulates as a warm pool around Indonesia. Water evaporating from the warm pool condenses to form huge clouds and heavy rains. Weather on the opposite side of the ocean is very different. The prevailing winds blow from the east, across the South American continent and before they reach the coastal belt they must cross the Andes, where they lose what remaining moisture they carry, then subside down the western side of the mountain range, warming by compression and becoming still drier. Consequently, the west coast of South America is extremely arid.

When the pressure distribution changes, weakening or reversing the winds, water from the warm pool flows eastward and accumulates above the water of the Peru Current. The flow of air also reverses, now approaching the coast from the west. The air is moist and it acquires more moisture as it crosses the warm surface water. When it reaches the coast and rises, its moisture condenses and falls as heavy rain. This is El Niño. The diagram on page 32 shows the difference between normal and El Niño conditions.

Historically, life was tough for the people living along the coast. They lived mainly by fishing, but the dry climate made it difficult to grow crops. Not surprisingly, they rejoiced when an El Niño delivered the rains that guaranteed the following year would be *un año de abundancia*, with enough food for everyone. The gift arrived at around Christmas, which is why they named it after the Christ child. Unfortunately, it does not bring glad tidings to everyone. As well as cold water, the many upwellings in the Peru Current carry to the surface nutrients scoured from the ocean floor. These feed a large plankton population which, in turn, provides food for vast shoals of fish, especially anchovies (*Engraulis ringens*, known as anchoveta), and the sea birds and seals that

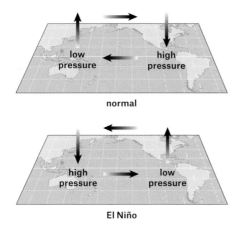

normal

El Niño

At intervals of two to seven years a change takes place in the South Pacific trade winds, linked to a change in the distribution of air pressure. The winds weaken, and sometimes reverse direction, and so does the wind-driven South Equatorial Current. Warm surface water moves eastward from the region of Indonesia and forms a pool off the South American coast. This reaches a peak in December and is known as El Niño.

feed on them. The seabird and seal droppings on coastal rocks have long been harvested as guano (the name means dung in Quechua), a valuable fertilizer rich in nitrogen, phosphorus, and potassium. The Peruvian fishery, based on anchovies, is huge, but during an El Niño the warm water suppresses the upwellings, the plankton disappear, the fish seek their food elsewhere, and the fishing communities suffer.

Indonesians and Malaysians also notice the change. With the disappearance of the warm pool the rate of evaporation slows, the rains cease, and often there is drought. The fires that swept across Sumatra in some recent years, causing serious air pollution over large parts of southeast Asia, were due to El Niño–induced droughts.

So El Niño can be quite dramatic, and its effects are felt more widely. As the map opposite shows, the dry area over Malaysia and Indonesia extends eastward halfway across the ocean. East Africa is also dry, as is northeast South America, while east Asia and western and northeastern North America enjoy a warmer winter than usual.

At other times, the pattern reverses. The pressure rises over the eastern South Pacific, falls in the west, and the winds and ocean current intensify. When that happens the already heavy rainfall in the west increases and there are floods, while the South American coastal belt becomes even more arid. And these effects, too, are felt elsewhere. Areas that were wetter under El Niño conditions become drier, drier areas become wetter, and warmer areas become cooler. For want of any other name, this intensification of normal conditions has come to be known as La Niña (female child). Because both are linked to the

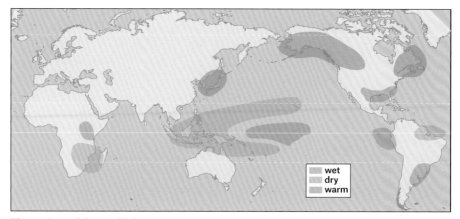

wet
dry
warm

The regions of the world that are warmer, drier, and wetter during an El Niño event.

southern oscillation, the complete cycle is known as an ENSO event (El Niño–southern oscillation).

These events have been traced over thousands of years. They vary in strength, but the one of 1997–98 was the strongest ever recorded and that of 1982–83 was also very strong. Some scientists have suggested on the basis of computer models that if global temperatures rise, ENSO events will become stronger and more frequent. There is no historical evidence to support this, however. These events have not grown stronger, more frequent, or lasted longer in records dating back to 1871, so there is no reason to suppose they will do so in years to come.

North Atlantic oscillation, Arctic oscillation

ENSO events do not spread their effects as far as Britain and western Europe, but that does not mean this part of the world escapes the far-reaching teleconnections triggered by major climate cycles. Winter storms, extreme winter cold, and unusually mild winters are often brought to us by one of the Northern Hemisphere's most important cycles, the North Atlantic Oscillation (NAO). The NAO exerts its effects throughout the year, but they are felt mainly in winter.

The high-latitude, subsiding side of the Hadley cells produces areas of semi-permanent high surface pressure, the subtropical highs. One of these is usually centred over the Azores, and is known as the Azores high, although sometimes it migrates to the western side of the ocean, when it is called the Bermuda high. There is also a semi-permanent area of low pressure centred over Iceland and known as the subpolar low or Iceland low. The NAO reflects

La Niña causes drought in parts of Australia and Indonesia, often leading to bush fires like the one shown here in Australia. Smoke from El Niño fires in Sumatra cause serious air pollution over a wide area of southern Asia.

the changing differences between the surface pressures in each of them. The extent of the difference is reported as an index, which can be positive or negative. The NAO index is positive when pressure is higher than usual in the Azores high and lower than usual in the Iceland low, so the difference between them is increased, and negative when the difference decreases, with pressure lower than usual in the Azores high and higher than usual in the Iceland low. The index increases and decreases from year to year and from time to time it switches between positive and negative. It can then remain in the same phase for several decades.

In the Northern Hemisphere, air circulates in a clockwise direction around centres of high pressure and anticlockwise around centres of low pressure. Consequently, air flows in an easterly direction on the southern side of the Iceland low and also on the northern side of the Azores high. This flow of air carries weather systems across the North Atlantic from west to east, but with a vigour that depends on the phase of the NAO index. As the illustration here shows, when the index is positive the two circulations extend farther and are stronger. During the negative phase they are weaker and further apart.

When the index is positive, there are more frequent and stronger winter storms that cross the ocean on more northerly tracks than usual. This brings mild, wet winters to western Europe and the eastern United States, and cold, dry winters to northern Canada and Greenland. During a negative phase there are fewer winter storms, they are weaker, and they cross the ocean on an approximately west–east track. Northern Europe then experiences a cold winter, and the winter is wet in the Mediterranean region. There are more cold periods and more snow than usual on the east coast of the United States, while Greenland has a mild winter.

Because the index often remains in a particular phase for several years, it can produce sequences of mild or cold, wet or dry winters. Sometimes these can persist for much longer. The index was strongly negative from about the mid 1950s until the 1970s, and, except for a brief negative period in the late 1970s, mainly positive from the 1970s until the late 1990s. Interestingly, the negative period coincided with a very slight fall in global mean temperatures and the latter, positive period with strong global warming. The index has been much more variable since 2000.

Extreme winter weather is also linked to a closely related but much less regular climate cycle,

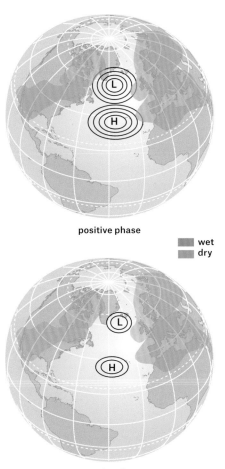

positive phase

■ wet
■ dry

negative phase

The North Atlantic Oscillation index goes through positive and negative phases. These have a strong influence on winter weather on both sides of the ocean.

the Arctic Oscillation (AO), also known as the Northern Annular Mode or Northern Hemisphere Annular Mode. This refers to the difference in surface pressure between those centred in the arctic and at about latitudes 37–45° N. When pressure is higher than usual in one it is usually lower in the other, and the phases change but with no discernible pattern.

When pressure is low in the arctic the AO index is said to be positive. This allows the polar front jet stream to blow strongly and approximately from west to east. The strong jet stream keeps arctic air enclosed in the arctic so it is unable to extend south. During the positive phase, storms crossing the North Atlantic follow a more northerly track, bringing wet weather to northern Europe and Alaska and dry weather to the western United States and the Mediterranean region. Greenland and Newfoundland are colder than usual and most of the United States is warmer.

When the AO is negative, conditions are reversed. Large waves develop in the track of the jet stream, allowing arctic air to penetrate southward, in winter often bringing extreme cold to most of the United States. The extreme cold and snow in North America and storms with heavy rain in Britain during the 2013–14 winter were due to a strongly negative AO.

Although the AO changes phase unpredictably, sometimes one sticks. It remained positive in most years from the 1970s to the mid 1990s, for instance, but then returned to oscillating between positive and negative.

Why the tropics are wet and the subtropics dry

Belém is a city in Brazil at latitude 1.45° S, almost on the equator, and 13 metres above sea level. Its geography ensures it has an equatorial climate with an average temperature of about 35°C (95°F) throughout the year. It is a rainy place, receiving an average 2440 mm (96.0 in.) of rain a year. Cardiff, in South Wales, also has a damp climate, but it receives a mere 1065 mm (41.9 in.) a year.

Both cities are on the coast, but the difference is that the sea off Cardiff is cool, while that off Belém and throughout the equatorial region is warm. In January 2014 it was at about 30°C (86°F) and the warm sea surface temperature with still warmer air above it generates a rate of evaporation that is much higher than is ever possible at Cardiff.

Liquid water is made from H_2O molecules that link together by hydrogen bonds between hydrogen atoms on one molecule and an oxygen atom on another, so they form small groups that move freely and can slide past one another. Hydrogen bonds are weak, so the groups are constantly breaking apart and reforming. The speed with which the groups move depends on their kinetic energy—the energy of motion. The warmer they are, the faster they go, which means the molecules in the sea off Belém whizz around much faster than those off Cardiff, where in summer the sea surface temperature struggles to reach about 15°C (59°F).

Water molecules that approach the surface with sufficient kinetic energy are able to break free from their groups and leap into the air. That is the process of evaporation, also called vaporization, and it marks the transformation of a liquid into a gas. Water evaporates much more readily from the equatorially warm water off Belém than from the cool waters of South Wales.

Once airborne, the equatorial water molecules are swept upward in the warm air rising on the low-latitude side of a Hadley cell. As it rises, the air with its load of water vapour expands and as it does so its molecules lose energy. The water molecules move more slowly until a point is reached where their kinetic energy is no longer sufficient to overcome the attraction of hydrogen for oxygen and the bonds start to reform. The molecules join into small groups and the small groups form minute droplets of liquid water. That is condensation.

Extra energy is needed to convert a liquid to a gas. This cools the sea a little, but when the water vapour condenses, the same amount of energy is released, as heat that warms the air. Being warmed, the air continues to rise. More water vapour condenses, releasing more warmth, and so the process continues. The condensed moisture forms cloud droplets and the clouds grow ever taller until they are storm clouds, some of them towering to a height of 16 kilometres. The droplets that compose them merge to form larger droplets, which grow until they are too heavy for the rising air to support them, so they fall as rain—often torrential rain. That is why Belém, like most places close to the equator, has such a rainy climate.

High above the surface, the air then moves away from the equator. It has lost much of its moisture and when it reaches about latitude 30° it subsides. As the air subsides, entering a region of higher pressure because of the greater weight of overlying air, it is compressed and its molecules gain energy, causing them to move faster. The temperature of the air rises, because the temperature of any substance is simply a measure of the speed with which its molecules move or vibrate. The air is already dry, but the amount of water vapour that air can carry is proportional to its temperature. The warmer the air, the more moisture it can carry as vapour, so as it subsides, the possibility of the condensation of such moisture as it still holds retreats, and the dry air becomes yet drier. It reaches the surface in the subtropics as warm, dry air that then flows outward, blocking the path of any moister air that might try to penetrate. That is

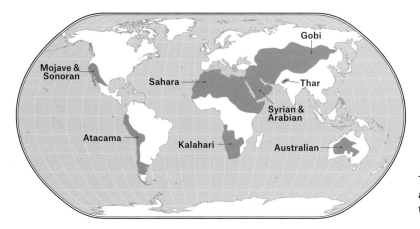

The world's major deserts. There are ten deserts in Australia, but this map shows them as one.

why there is a belt of deserts around the subtropics of both hemispheres—the Sahara, Arabian, and Syrian Deserts of the Northern Hemisphere, and the Kalahari, Great Victoria, Great Sandy, Gibson, and Simpson Deserts of the Southern Hemisphere.

Types of deserts

For many of us the word *desert* conjures an image of high sand dunes stretching into the distance. Such deserts exist, of course, especially in the Sahara, which is the world's largest desert covering about nine million square kilometres, and where there are seas of sand called ergs, from the Arabic *areg*, meaning ocean. These are located in the northern part of the desert. The Grand Erg Oriental covers about 100,000 km^2 and the Grand Erg Occidental about 70,000 km^2. Vast though they are, the ergs occupy barely 2% of the total area. The remainder of the surface consists of exposed bedrock and large boulders called hammada, and layers of gravel and pebbles called reg. The name of the desert is Arabic. *Sahrá* means wilderness.

The Sahara is a subtropical desert that extends eastward as the Arabian and Syrian Deserts. As the map shows, however, there are deserts in every continent, not only in the subtropics. In fact, deserts cover almost half of the world's land area. So what does it take for a desert to form?

The moment raindrops fall from a cloud they enter unsaturated air and begin to evaporate. Sometimes all of them vaporize before reaching the ground, so a veil of falling rain hangs beneath the cloud base but extends only partway to the surface. It is called virga or fallstreaks and is a familiar sight.

Despite being renowned for its oceans of sand, much of the Sahara is hammada, a surface of bare rock, stones, and gravel.

Desert windstorms can create dust or sand storms that drive people to seek shelter. The storm can reduce visibility to almost zero and the sand and dust penetrate clothing and enter buildings through the tiniest gaps.

Virga (fallstreaks) forms a veil below the base of a cloud. It consists of rain that evaporates before reaching the surface.

Rain that survives long enough to reach the ground enters small crevices and pores between mineral particles, but unless the air is saturated the water continues to evaporate, and much of the moisture that is absorbed by plant roots returns to the air by transpiration. Evaporation and transpiration are difficult to separate for measurement and are often considered together and called evapotranspiration. During dry spells evapotranspiration removes all of the water that reaches the ground, so the ground becomes progressively drier, but in most climates the dry spells do not last and eventually the lost moisture is restored. That is not what happens in a desert.

Suppose that the supply of water is limitless. The amount that evapotranspiration would remove in such a situation is called the potential evapotranspiration. It is the amount of water evapotranspiration is capable of removing. If the amount of rainfall, averaged over the year, is less than the potential evapotranspiration, a desert will form. Most deserts experience occasional rain, often torrential rain, but invariably the dry weather that follows continues long enough to vaporize all of it. Water evaporates more rapidly into warm air than into cold air, so potential evapotranspiration varies with latitude, but

The Atacama, a west coast desert in Chile, is one of the world's most arid deserts.

anywhere in the world a desert will develop if the average annual rainfall is less than 250 mm (9.8 in.). Areas that satisfy this requirement are not confined to the subtropics.

The Atacama Desert lies along the west coast of Chile. Iquique, a town in the north, once went four years with no rain at all and in the fifth year a single shower delivered 15 mm (0.6 in.). Yet Iquique is a coastal town. The desert as a whole receives an average 10 mm (0.4 in.) of precipitation a year and most of it arrives not as rain, but fog. The Namib Desert, in Namibia, lies on the west coast of Africa and is the African equivalent of the Atacama, though not quite so dry with an annual precipitation of about 50 mm (2.0 in.). Again, some of that precipitation is in the form of fog.

These are west coast deserts lying in the tropics. Their prevailing winds are from the east and must cross a continent before reaching them, but that is only

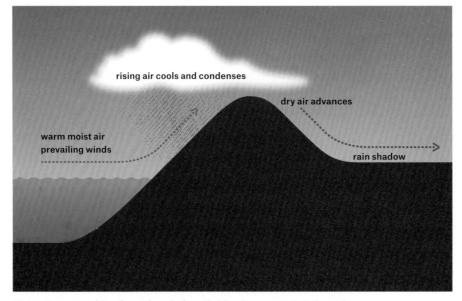

Moist air approaching from the windward side of a mountain is forced to rise. Its temperature falls and water vapour condenses, forming cloud and delivering precipitation. On the lee side the dry subsiding air warms by compression, further lowering its relative humidity.

part of the reason for their aridity. By day the desert surface heats rapidly and by midmorning warm air is rising vigorously and drawing in air from over the sea to replace it. This is a sea breeze and might be expected to carry moist air, but offshore there is a cool eastern boundary current—the Peru Current off Chile and the Benguela Current off Namibia. Moist air that is drawn toward land is chilled as it crosses the current and its water vapour condenses to form fog and low cloud, but these are not deep or dense enough to form rain. The deserts are often foggy and at night some of the fog condenses on cool surfaces, but the boundary currents contribute to the aridity.

Other deserts form in the rain shadow of mountain ranges. Most of the North American deserts are of this type, as is the Thar or Great Indian Desert, and there are many more. The Mojave Desert, for instance, receives about 125 mm (4.9 in.) of precipitation a year, some of it as snow and frost; the temperature on winter nights often falls below freezing. Rain shadow deserts form on the lee side of mountains. Moist air approaching on the windward side is forced to rise. It expands and cools, and its water vapour condenses to form cloud and precipitation. As the air subsides on the lee side of the barrier it warms by compression, which further reduces its relative humidity. The diagram above illustrates the process.

Desert hyacinth (*Cistanche tubulosa*) is widely distributed in the deserts of Arabia and the Middle East. The plant parasitizes the roots of other plants and flies pollinate its flowers.

The Gobi Desert, in Central Asia, is thousands of miles from the sea.
It receives an average 50 mm (2 in.) of precipitation a year.

Eurasia is a vast continent and there are regions at its heart so far from the ocean that the air reaching them has lost most of its moisture before it arrives, and to compound their problems these areas are surrounded by mountains, so they lie in rain shadows. The centre of the Gobi Desert receives up to 50 mm (2.0 in.) of precipitation a year and the driest part of the Takla Makan Desert to its west receives an average 10 mm (0.4 in.).

One of the driest of all deserts lies beneath an average 2 kilometres of ice, which seems paradoxical, but is due to the fact that the temperature never rises sufficiently to melt snow, so the snow that falls remains to accumulate year by year for millions of years. The centre of Antarctica receives less than 50 mm (2.0 in.) of rainfall-equivalent a year and only the coastal fringes receive more than 200 mm (7.9 in.). The precipitation falls as snow, of course, but the consistency of snow varies greatly, so it is always melted and its depth reported as rainfall-equivalent in order to produce comparable measurements. The interior of Greenland is also very dry, with an average annual rainfall-equivalent precipitation in some places of less than 200 mm (7.9 in.) and of only 120 mm (4.7 in.) at Thule.

Flags of the nations mark the location of the South Pole, where the only inhabitants are research scientists, four of whom are seen towing sledges containing supplies and equipment. This is the coldest place on earth, and also the driest.

There are several reasons for the aridity of polar regions. The first is the extreme cold. The amount of water vapour air can hold is proportional to the temperature. The colder the air, the less moisture it can carry and the temperatures over Antarctica and central Greenland are so low that the air can hold very little moisture. The air itself is extremely dry. The low temperature is due to the geographical location, but not only that. The thickness of the Greenland ice sheet averages about 1500 metres, raising the surface high above sea level, and the Antarctic ice sheet is at an even higher elevation. The rock beneath its ice is an average 2.44 kilometres above sea level and the ice sheet raises the surface above 4 kilometres.

Already dry because of the temperature and elevation, the Greenland and Antarctic winds block the entry of moist air. Cold air is dense and subsides, then flows outward. Ice sheets are not level. They form domes, highest near the centre, and dense, subsiding air accelerates downhill, producing fierce winds near the coast. The dense air moves beneath moister, less dense air, raising it aloft so it loses its moisture, and preventing it from penetrating the interior.

Structure and composition of the atmosphere

If you've ever climbed a mountain you'll know that the higher you trudge the colder and thinner the air will be. Climb high enough and breathing becomes difficult. You may need additional oxygen. The density of the air decreases quite rapidly with altitude. The table shown here illustrates this.

Half of the all the air in the entire atmosphere is in the lowest 6 kilometres and 90% of it lies below about 16 kilometres, yet the atmosphere extends to an altitude of more than 500 kilometres, beyond which it merges with the outer fringes of the solar atmosphere and the gases in interplanetary space. The rapid decrease in density is due to the fact that air is highly compressible—think of car tyres—and the weight of the overlying atmosphere compresses it.

Weather forecasts for climbers report the expected air temperature at a specified height. Temperature continues to decrease with height until you reach an altitude of about 16 kilometres above sea level at the equator, 11 kilometres in middle latitudes, and 8 kilometres at the North and South Poles. This is the height of the tropopause and above it the temperature remains constant with height up to about 20 kilometres.

At this height the average temperature is –65°C (–85°F) over the equator, –55°C (–67°F) over middle latitudes, and –30°C (–22°F) over the poles, which sounds distinctly odd, because everybody knows that equatorial air is much warmer than polar air. The temperature difference is due to the difference in the average height of the tropopause, which is determined by the vigour of the convective movements in the air. Convection is most vigorous over the equator and least vigorous over the poles, so the tropopause is highest over the equator, and temperature decreases with height all the way to the tropopause, so the higher the tropopause the lower the temperature there will be.

Cold air subsides because it is relatively dense and pushes beneath less dense warmer air, raising it aloft. That is why warm air rises—it is not defying gravity! When the rising air enters a region where the surrounding air is at the same density and temperature, it can rise no farther. So a layer of air in which the temperature and density remain constant with height forms a barrier to

ATMOSPHERIC DENSITY AND HEIGHT

Height (km)	Density (kg/m3)
30	0.02
25	0.04
20	0.09
19	0.10
18	0.12
17	0.14
16	0.17
15	0.20
14	0.23
13	0.27
12	0.31
11	0.37
10	0.41
9	0.47
8	0.53
7	0.59
6	0.66
5	0.74
4	0.82
3	0.91
2	1.01
1	1.11
0	1.23

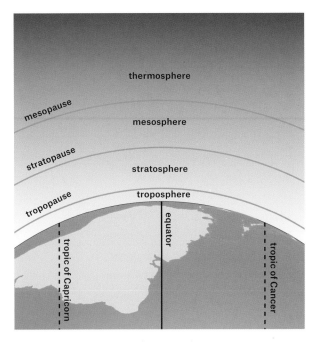

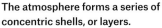

The atmosphere forms a series of concentric shells, or layers.

rising air. And that is the significance of the tropopause. It is a thermal and density barrier to rising air.

The troposphere is the region of the atmosphere lying beneath the tropopause, in which the constant movement of air ensures that it is thoroughly mixed. It is the part of the atmosphere where weather phenomena occur. The existence of an upper boundary suggests a region of the atmosphere above the tropopause and, indeed, the troposphere is the lowest of several atmospheric layers that form a series of concentric spheres blanketing the Earth. The troposphere is sometimes called the lower atmosphere, and all the air above it comprises the upper atmosphere.

Above the tropopause lies the stratosphere. At first the temperature remains constant with height, but above about 20 kilometres it begins to increase because at this height, oxygen and ozone are starting to absorb ultraviolet sunlight. Above about 32 kilometres, the warming accelerates until at the stratopause it sometimes reaches 27°C (81°F). At that height the atmospheric pressure averages 100 pascals, which is one-thousandth of the sea-level pressure.

The stratopause is the upper boundary of the stratosphere. In summer it is at about 55 kilometres over the equator and poles and about 50 kilometres over middle latitudes. In winter it is at about 50 kilometres over the equator and 60

COMPOSITION OF THE ATMOSPHERE

Gas	Chemical formula	Abundance
MAJOR CONSTITUENTS		
nitrogen	N_2	78.08%
oxygen	O_2	20.95%
argon	Ar	0.93%
water vapour	H_2O	variable
MINOR CONSTITUENTS		
carbon dioxide	CO_2	396 ppmv
neon	Ne	18 ppmv
helium	He	5 ppmv
methane	CH_4	1.8 ppmv
krypton	Kr	1 ppmv
hydrogen	H_2	0.5 ppmv
nitrous oxide	N_2O	0.3 ppmv
carbon monoxide	CO	0.05–0.2 ppmv
xenon	Xe	0.08 ppmv
ozone	O_3	variable
TRACE CONSTITUENTS		
ammonia	NH_3	4 ppbv
nitrogen dioxide	NO_2	1 ppbv
sulphur dioxide	SO_2	1 ppbv
hydrogen sulphide	H_2S	0.05 ppbv

ppmv = parts per million by volume
ppbv = parts per billion by volume

kilometres over the poles. The temperature through the stratopause remains constant with increasing height and the layer above it is the mesosphere, where the temperature falls sharply with height to about –90°C (–130°F) in winter but occasionally –30°C (–22°F) in summer. The mesopause is the upper boundary of the mesosphere and above it the thermosphere extends to the thermopause at 500–1000 kilometres, where the temperature sometimes exceeds 1000°C (1832°F). The rise in temperature is due to the absorption of ultraviolet light by oxygen atoms, but the atmosphere in the thermosphere is so tenuous that it does not warm orbiting satellites. The diagram on page 47 shows the vertical structure of the atmosphere.

The chemical composition of the air is constant throughout the troposphere and stratosphere because of convective mixing. This ceases above about 100 kilometres and there the gases form layers, the lighter ones lying on top of the heavier ones. Above about 120 kilometres ultraviolet radiation breaks oxygen molecules apart (O_2 + photon UV → O + O) and more than half the oxygen is present in the atomic form.

The table opposite shows the chemical composition of the troposphere and stratosphere.

The "greenhouse effect"

Sunshine warms the land and sea surface, the warmed surfaces radiate their warmth back into space, and the amounts of incoming and outgoing radiation are approximately in balance. If that were not so, the world would grow steadily warmer or cooler. Of course, sometimes the world is warmer or cooler than it is at other times, and the atmospheric temperature increased during the latter part of the twentieth century, but over long periods of millions of years the energy budget balances.

The balance is not immediate, however. If Earth lost its warmth as fast as it receives it, the climate would be a good deal chillier, with an average surface temperature between about –15°C (5°F) and –25°C (–13°F). Most or all of the oceans would be covered by ice and the dry land would be permanently frozen and extremely dry. In fact, the average global surface temperature is about 14°C (57°F), the oceans are frozen over only near the poles, and except in the deserts precipitation waters the land.

We live in happier climes because the departure of the Earth's outgoing radiation is delayed. Nitrogen and oxygen, which between them make up 99% of the air, are transparent to radiation at the wavelengths of sunlight, 0.2–4.0 μm. The Earth radiates at 5.0–50.0 μm with a strong peak at about 12 μm, and the atmosphere is partly opaque to radiation in this waveband. Water vapour absorbs radiation at 5.3–7.7 μm and at all wavelengths above 20 μm, carbon dioxide absorbs at 13.1–16.9 μm, and ozone absorbs at 9.4–9.8 μm. Methane, nitrous oxide, hydrofluorcarbon (HFC) and chlorofluorocarbon (CFC) compounds, and sulphur hexafluoride absorb at other wavelengths. No gas absorbs at 8.5–13.0 μm, a waveband that forms a window through which outgoing radiation can escape to space. Significantly, it covers the radiation peak at 12 μm.

When molecules of these gases absorb radiation they immediately re-radiate it at a longer wavelength and in all directions. Some of this radiation travels upward, some to the sides, and some downward, and as it passes through the air some of it strikes other gas molecules and is absorbed, so the process continues. The overall effect is that the outgoing radiation becomes trapped by the atmospheric gases, raising the air temperature by 30–40°C (54–72°F). All the radiation at the window wavelength escapes and a balance develops between the radiation being trapped and the radiation escaping, which prevents freezing or overheating.

Several nineteenth-century physicists studied the way radiation passes through air and noted that certain gases absorb at particular wavelengths. In 1917, Alexander Graham Bell is said to have written that the unchecked burning of fossil fuels would have a sort of greenhouse effect and that the greenhouse would become a sort of hothouse. Thus the effect came to be known as the greenhouse effect.

The name is misleading, though, because this is not at all what happens inside a greenhouse. The transparent windows of a greenhouse allow sunshine to enter and warm exposed surfaces, and contact with those surfaces warms the air. The air temperature continues to rise, however, not because radiation is being absorbed and re-radiated, but because the greenhouse is closed and the warm air cannot escape. Open the vents or the door and the temperature quickly starts to fall.

Since about the middle of the nineteenth century, manufacturing industry, and more recently transport, has used coal, petroleum, and gas as a source of primary energy. Burning these so-called fossil fuels releases carbon dioxide as a by-product and the atmospheric concentration of this gas has been increasing at an accelerating rate as industrialization expands globally. Forest clearance and the kilning of limestone to make cement also release carbon dioxide. Atmospheric concentrations of methane and some other "greenhouse gases" have also increased over the past century.

During the same period, the global mean temperature has risen, by about

Greenhouses or glasshouses have been used at least since the thirteenth century to grow crops out of season or isolated from climates in which they could not survive.

0.7°C (1.26°F) since 1880 of which 0.5°C (0.9°F) occurred between 1976 and 1996. The rise was most pronounced in the early part of the century, reaching a peak in about 1940, then the mean temperature remained constant or decreased very slightly through the 1940s, 1950s, 1960s, and early 1970s before starting to rise again in 1976. There has been no statistically significant change in the temperature since January 1997.

Part of the twentieth-century warming was due to the climate's recovery from the Little Ice Age, but a proportion is likely to have been due to the increase in atmospheric concentration of greenhouse gases. The size of the contribution these have made is highly controversial and although computer models predict that the temperature will continue to rise, the same models have consistently overestimated the rate of temperature change and failed to predict the halt in the rise, so their predictions are unreliable. To distinguish the warming due to our release of greenhouse gases from the natural effect, it is sometimes called the enhanced greenhouse effect.

Air masses and fronts

Like many explanations, it's perfectly obvious once someone else has spelled it out. After a while, air sitting over a continent will lose much of its moisture and become dry, air sitting over an ocean will become moist, tropical air will be warm, and polar air cold. What is less obvious is that the physical forces acting on these large bodies of air will cause them to drift away, so they are already in motion as they acquire their characteristics.

Weather systems cover very large areas. Every TV weather forecast affirms this when the forecaster tells of weather approaching from the other side of an ocean or continent. It was not until the middle of the nineteenth century that it was possible to know this. That is when the telegraph was invented. Until then, information could travel no faster than a messenger on a galloping horse could carry it. The telegraph allowed meteorologists to gather data rapidly from widely scattered observers and when they plotted the data on maps, patterns began to emerge. It took time, of course, because at first the telegraph lines followed rail lines leaving large areas without access to them, and observers had to be recruited, trained, and provided with instruments.

In 1917 a geophysical institute was established in Bergen, Norway, to take over atmospheric research previously conducted by the Bergen Museum and the Biological Station. In the years that followed, a team of scientists at the Bergen Geophysical Institute established a network of observers who sent them meteorological data from all across Scandinavia. When they plotted on charts the instrument readings for temperature and humidity, all taken at the same pre-arranged times, they noted that these values remained constant over large areas, and since air density can be inferred from the temperature, density was also constant. They had identified vast volumes of essentially homogenous air, and they decided to call them air masses. An air mass is a body of air in which the physical attributes are approximately constant throughout and at every level from the surface to the tropopause.

Since then air masses have been identified across the whole world, and classified. The classification is very simple. First, air masses acquire their characteristics either over continents or over oceans, so they are labelled continental (c) or maritime (m). Their temperatures depend on the latitude in which they form, so they are identified as arctic (A), polar (P), tropical (T), or equatorial (E). These designations combine to give seven types of air mass: continental arctic (cA), continental polar (cP), continental tropical (cT), maritime arctic (mA), maritime polar (mP), maritime tropical (mT), and maritime equatorial (mE). There is no continental equatorial air mass, because there is no

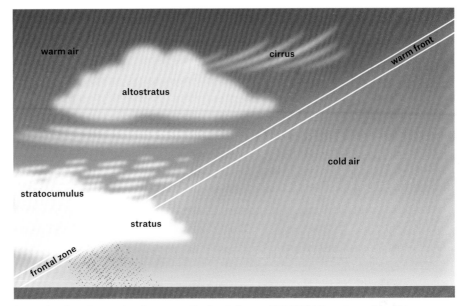

A weather front is a region, known as the frontal zone, 100–200 kilometres wide. This is a warm front, with air behind the front warmer than the air ahead of the front, and it extends at an angle of 0.5–1.0 degrees (here much exaggerated) from the surface to the tropopause. The diagram also shows the types of cloud associated with a warm front.

continental landmass in the equatorial region. Some scientists distinguish monsoon air, but it is indistinguishable from mT air.

The scientists of what came to be known as the Bergen School observed something else. Adjacent air masses were of different densities and fluids of different densities do not mix readily, because the denser sinks beneath the less dense. It is simple to demonstrate this by heating water containing a dye and then pouring it very carefully into a container of clear cold water. The coloured water will float on top of the clear water.

Repeated observations confirmed that air masses move, dragged along by the prevailing upper-level winds. In middle latitudes they travel in a generally easterly direction, but not all at the same speed. Cold air, being denser, moves faster, at an average 35 km/h, than warm air moving at 24 km/h. The cold air pushes against the warm air, but without mixing. To the meteorologists, the boundary between cold and warm air suggested a front between opposing armies. They called such boundaries fronts, identified by the change in air temperature as the front passes. If the air behind the front is cooler than the air ahead of it, it is a cold front, and if the air is warmer it is a warm front.

Although a front is shown on a map as a thin line, in fact it is a zone 100–200 kilometres wide through which the characteristics change rapidly. The diagram on page 53 shows a weather front in three dimensions, extending from the surface to the tropopause. As the diagram indicates, fronts slope, although the slope in the diagram is greatly exaggerated in order to fit it on the page. A warm front rises at an angle of 0.5–1.0 degrees to the surface. Cold air, being denser, is more strongly affected by friction with the surface. Consequently, it is slowed at the surface but not at higher levels, so the upper air moves faster than the surface air making the front slope more steeply, at about 2 degrees. Both slopes are shallow. The top of a warm front, at the level of the tropopause, is 550–1200 kilometres from its position on the surface, and the top of a cold front is about 300 kilometres away. Cloud begins to appear and precipitation falls from it long before an advancing front arrives and persists long after the front has passed.

As a cold front passes the surface air pressure rises, the temperature falls, and the wind direction changes, in the Northern Hemisphere veering, which means the direction changes to the right, from southerly to westerly, for example. Clouds form along the front and bring showers that can fall up to 200 kilometres ahead of or behind the position of the front on the surface. If the warmer air ahead of the front is very moist, the showers can include thunderstorms. If the air is dry, on the other hand, the front may produce nothing more dramatic than a few light showers or no cloud or precipitation at all. The cold air behind the front is usually stable. It is subsiding and brings clear skies.

A warm front brings a rise in temperature and a fall in pressure. The sky becomes overcast, with very high clouds appearing first, and precipitation starts falling up to 400 kilometres ahead of the arrival of the front at surface level. If the warm air is dry, there may be little cloud or precipitation, but if it is moist the precipitation may be prolonged and steady. The weather behind the front is usually drier and warmer.

Continental climates and maritime climates

On a hot summer afternoon the sand on a beach can be so warm that it hurts your bare feet and you have to run to cross it. Yet when you reach the sea it feels cool or even cold. The Sun beats down equally on both, and they are adjacent, so how can this be? The answer is that different substances must absorb different amounts of energy before their temperatures rise. This is known as the heat capacity (or thermal capacity) of each substance and it is a fundamental property of all substances.

The point is that water has a very high heat capacity, which means that a large amount of energy is needed to make its temperature rise. To be precise, it takes 4.19 joules of energy to raise the temperature of 1 gram of fresh water by 1°C (1.8°F), and 3.93 joules to heat seawater by 1°C. It takes only 0.84 joules to raise the temperature of 1 gram of sand by 1°C. Heat capacity varies a little with the temperature, so it is usual to specify this. In the case of fresh water it is 15°C (59°F), of seawater 17°C (63°F), and of sand from 20–100°C (68–212°F). And that is why running into the sea will cool your feet that were scorched by the sand.

Sand and all kinds of rock heat up much more quickly than water does, and they also lose their heat much faster. This means that dry ground warms quickly in spring, becomes very warm in summer, cools rapidly in autumn, and is very cold in winter. Water, by contrast, takes a long time to warm and to cool. Hot, dry, continental air that crosses the ocean in summer is cooled by contact with water that is still warming, and cold continental air that crosses in winter is warmed by contact with water that is still cooling. The result is that places in the interior of continents have drier weather than places near the coast, and are hotter in summer and colder in winter. The brave souls who leap into the sea on New Year's Day are entering water that may be warmer than the air temperature.

Edmonton, Alberta, and Dublin, Ireland, are both at latitude 53° N. In winter (December–February), the mean day temperature at Edmonton is –7°C (19°F) and the night temperature –17°C (1°F). In Dublin, the mean day temperature is 8°C (46°F) and night temperature 2°C (36°F). So Dublin has much milder winters than Edmonton. In summer (June–August) the day and night mean temperatures in Dublin are 19°C (66°F) and 10°C (50°F), and in Edmonton they are 22°C (72°F) and 8°C (46°F), so Edmonton has warmer summers. The temperature range, which is the difference between the highest and lowest mean temperature, is 43°C (77°F) in Edmonton and 19°C (34°F) in Dublin. Edmonton has a far more extreme climate. It is also much drier, with an annual precipitation of 440 mm (17.3 in.) compared with 760 mm (29.9 in.) in Dublin.

The difference between the two climates is due entirely to their locations and can be expressed as their continentality, which is the extent to which each climate departs from the most extreme continental type. Occasionally, continental air reaches Dublin and the effect of maritime air can extend deep into a continental interior, so the concept of continentality is not precise. Nevertheless, it provides a useful thumbnail impression of the change in the weather you may expect if you move to or far away from the coast of an ocean, and the

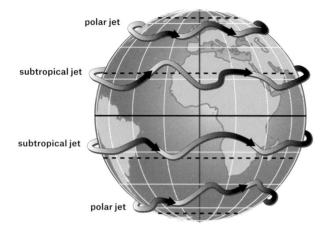

The polar jet stream strongly affects weather conditions over middle latitudes. The subtropical jet has less effect on surface weather.

degree of continentality can be calculated as a percentage. The calculation reveals that the degree of continentality at Edmonton is 93% and that at Dublin is 39%.

Jet streams and why they wander

During World War II, military aircraft routinely flew at altitudes higher than aircraft had flown before the war, and navigators on American aircraft flying over the North Pacific and German navigators flying over the Mediterranean made a curious discovery. On some journeys, but by no means all of them, they found their flying times were drastically increased or decreased by a high-level wind blowing at 100 km/h and often more. They were not the first people to observe these ribbons of wind, but they were the first to encounter them repeatedly. They called them jet streams. Typically a jet stream is hundreds of kilometres wide, often no more than 5 kilometres deep, and thousands of kilometres long, and it is found close to the tropopause.

There are several jet streams, the most important being the polar front and subtropical jet streams. The polar front jet stream blows at a height of 7–12 kilometres, in winter between latitudes 30° N and 40° N and in summer between 40° N and 50° N. In the Southern Hemisphere it blows at about 50° S in winter and 45° S in summer. The subtropical jet stream blows at about 30° N and S

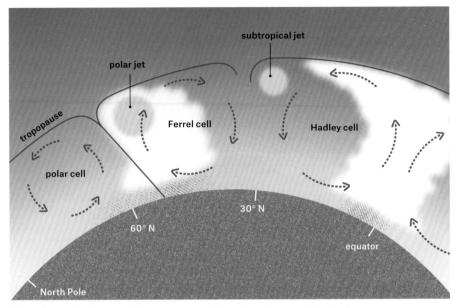

The polar and subtropical jet streams are situated at the top of fronts, the subtropical jet on the poleward side of the Hadley cells and the polar front jet on the poleward side of the Ferrel cells.

throughout the year. It is more persistent than the polar front jet stream, but weaker and also higher, blowing at 10–16 kilometres. The illustration opposite shows their approximate locations.

Both jet streams are located at the top of a major front. The polar front is the boundary between tropical and polar air where air is rising in the Ferrel and polar cells, and the subtropical front is between the Hadley and Ferrel cells. The diagram above shows this in cross section. These fronts are between cold air on the poleward side and warm air on the equatorward side. This is highly significant in middle latitudes, because it means places on the surface are exposed to polar or tropical air depending on which side they are of the polar front jet stream, and if the jet stream moves north or south—which it does from time to time—the weather will change dramatically. The extreme weather of the 2013–14 winter, when North America suffered arctic temperatures and heavy snow while western Europe experienced mild temperatures but persistent heavy rain that caused flooding, was due entirely to loops in the polar front jet stream that took it far to the south over North America and far to the north over Europe.

Jet streams are thermal winds. That is to say, they are generated by a large

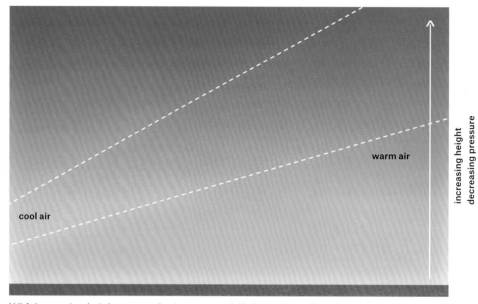

increasing height
decreasing pressure

warm air

cool air

With increasing height, atmospheric pressure falls faster in cool air than it does in warm air, so a surface of even pressure slopes upward from the cool to the warm air.

temperature difference over a short horizontal distance. There is a change in temperature across all fronts, but most fronts are fairly short-lived, whereas the polar and subtropical fronts are permanent features of the atmospheric circulation. At these fronts, the temperature gradient is associated with a horizontal pressure gradient that increases with height, reaching a maximum close to the tropopause.

Cold air is denser than warm air, and air is highly compressible. Consequently, air pressure decreases with increasing height faster in a column of cold air than it does in an adjacent column of warm air. At any given height, therefore, a surface where the pressure is the same throughout (an isobaric surface) will slope upward from the cold to the warm, and, as the illustration above shows, that slope will increase with increasing height. This means that the horizontal pressure gradient increases with increasing height and since the wind speed is proportional to the pressure gradient, the wind speed also increases with height.

At the top of the front, however, something else happens. As the cross section diagram shows, the frontal slope means the warm air lies over the top of the cold air, intensifying the gradient still more, and forming the ribbon of fast-moving air that is the jet stream. The wind blows parallel to the gradient

Chicago, Illinois, in the deep freeze of December 2013 and January 2014.

because of the Coriolis effect, with the cold air to its left in the Northern Hemisphere and to its right in the Southern Hemisphere, producing a westerly wind in both hemispheres.

So we start with fronts and their jet streams that are aligned parallel to lines of latitude. That, alas, is not how they remain, because the Rocky Mountains interfere. The Rockies are aligned approximately north–south and not only the polar front jet stream has to cross them, but also all the lesser winds below it. The air is forced upward, but also deflected horizontally, so the airflow, including the jet stream, moves into a higher or lower latitude. Friction with the adjacent air combined with the effect of the Earth's rotation then deflects the stream in the opposite direction. The magnitude of the force causing the deflection, known as planetary vorticity, increases with latitude. The airstream overshoots, and the change in magnitude of planetary vorticity changes, turning the stream back again. This generates a series of waves, known as Rossby waves after Carl-Gustaf Rossby, the Swedish-American meteorologist who first identified them. Rossby waves propagate from east to west in both hemispheres at an average speed of about 60° of longitude a week, and have a wavelength of 4000–6000 kilometres. Their amplitude—the distance between the crest of one wave and the trough of the next—varies, over a period usually of three to eight weeks. This variation is highly irregular, however. It is measured as a zonal index, based on the difference in pressure between latitudes 33° N and 55° N.

Why midlatitude weather is so variable

Air is drawn into jet streams and expelled from them. Air converges (is drawn together into a tighter stream) as it is drawn into the core of the jet stream, where the wind speed is greatest, and air that leaves the core diverges (forms a wider stream). That is because air being drawn inward accelerates, and an accelerating flow becomes narrower, while an airstream flowing outward decelerates and widens.

Convergence causes the air pressure to increase and that forces air out of the high-pressure region. This air subsides down the front all the way to the surface, where it produces an area of high pressure. Air that diverges as it is expelled from the jet stream produces a region of low pressure that draws air upward to fill it, thereby forming an area of low pressure at the surface. These areas of surface high and low pressure produce weather systems and the generally eastward flow of the jet stream drags them from west to east.

An area of low pressure is known as a cyclone, and if it forms outside the tropics in association with weather fronts, it is often called a depression. A depression may be too weak to produce much weather, but if it does generate weather, this will be mild and with persistent drizzle or light rain.

The circulation of air around the developing cyclone produces a small undulation in the polar front. The undulation grows into a wave with a clearly defined crest, and a region of warm air, called the warm sector, becomes enclosed between a warm and cold front. The air flows anticlockwise (in the Northern Hemisphere) around the wave crest, and the depression becomes a frontal cyclone. Cold air moves faster than warm air. It pushes beneath the warm air, raising it aloft, and the cold front starts to override the warm front, making the wave crest steeper. As more of the warm air is raised aloft the two fronts occlude. When all the warm air is clear of the surface the fronts dissolve and the original front reappears. The diagram opposite shows the sequence of events in the life cycle of a frontal system.

Frontal depressions tend to follow one another in families. No sooner has one system passed than another approaches from the west in a dreary procession.

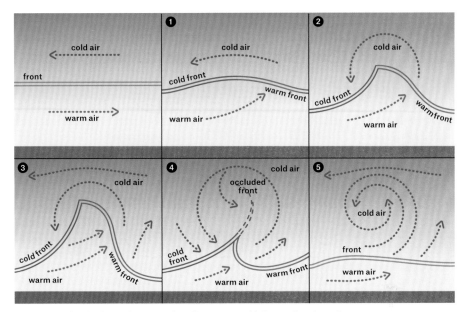

The life cycle of a frontal system has five stages. (1) Fronts begin to form when an undulation develops in the front. (2) This grows into a wave with a marked crest where the warm and cold fronts join. (3) Cold air advances beneath the warm air and the cold front starts to override the warm front. (4) The fronts occlude as the cold air raises the warm air clear of the surface. (5) When all the warm air has been raised aloft, the system dissolves and the original front reappears.

Meteorologists can see them coming, partly through monitoring the distribution of atmospheric pressure and partly through satellite images that reveal tell-tale cloud patterns. The clouds also reveal how vigorous the frontal systems are and the weather they will bring. Forecasting just where and when they will arrive is a good deal trickier, however. The systems move with the jet stream and the jet stream is inherently unstable. It flaps this way and that like a flag in the wind (but more slowly) so approaching weather systems often change track. An individual garden, even a huge one, is not much of a target and the system may well skirt it to the north or south, just as a system forecast to do exactly that may change track and hit the garden square on. Weather in middle latitudes is notoriously difficult to forecast.

It gets worse, because every so often the weather stops moving altogether. Fronts sometimes remain stationary for several days so a particular type of weather persists for a while, but sometimes it persists for much longer, because

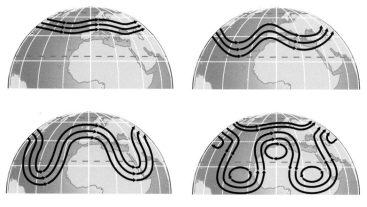

Waves in the polar front jet stream (1) become more pronounced (2) and eventually extreme (3). The jet stream partly breaks down (4) with isolated anticyclones to the north and cyclones to the south.

Northern Hemisphere

something different has happened. Approaching frontal systems divert to the north or south taking their weather with them. This produces settled weather that may last for a week or two but occasionally for long enough to cause a drought or floods, a scorching heat wave in summer or a deep freeze in winter. Something is blocking the approaching weather, pushing it to the side.

Once again the jet stream is to blame. To be more precise, it is the Rossby waves that lead to blocking. Remember that planetary vorticity increases in strength with increasing distance from the equator. Vorticity is the tendency for a fluid moving across the Earth's moving surface to follow a circular path around an axis at right angles to its direction of flow. When the jet stream turns poleward, planetary vorticity increases, turning the wind toward the equator. As it turns equatorward and overshoots, planetary vorticity decreases and the wind turns poleward again. Thus the Rossby waves begin as gentle undulations but gradually become more and more pronounced. The waves pointing toward the equator are called ridges and those pointing toward the pole are troughs. Ridges bring high pressure and polar air into low latitudes and troughs bring tropical air and low pressure into high latitudes. The diagram above shows how this pattern develops.

Eventually the undulations are so extreme that the wind takes a short cut across the narrowest necks in the ridges and troughs. This produces isolated cells. In the Northern Hemisphere the air flows clockwise—anticyclonically— around the cells in the north and anticlockwise—cyclonically—around the cells to the south. The anticyclones are areas of high surface pressure and the cyclones areas of low surface pressure and this pattern continues until the cyclones fill and the anticyclones weaken and the jet stream resumes its easterly track. The anticyclones bring fine weather, hot in summer and bitterly cold in winter, with sunshine and light winds. The cyclones are depressions that bring rain and make it difficult to remember that it is the pressure that is depressed, not the people enduring it.

When air moves approximately parallel to the equator its flow is said to be zonal, and when it flows parallel to the lines of longitude (meridians) the flow is meridional. During the development of the Rossby waves the jet stream flow, which is originally zonal, becomes increasing meridional. The jet stream is driven by a steep pressure gradient, with high pressure on the poleward side and low pressure on the equatorward side. The jet stream flow is strongly zonal when the pressure gradient is at its steepest. As the Rossby waves develop, the pressure gradient weakens and the flow becomes increasingly meridional. The changing strength of the pressure gradient and jet stream is measured as a zonal index, calculated from the difference in pressure on either side of the polar front, usually in latitudes 33° and 55°, and the growth and final breakdown of the Rossby waves represents a cyclical change in the zonal index, called the index cycle.

How climates are classified

Plymouth, in the south of England, has a very different climate from Winnipeg, and both climates differ from that of Bangkok. Climates vary widely and because we are impelled to categorize things, we feel a need to arrange climates into types, to pigeonhole them. There is nothing new in this. It began in ancient Greece. The Greeks divided the world's climates into frigid, which were impossibly cold; torrid, which were impossibly hot; and between them the temperate climates where people (Greeks) lived. We've abandoned frigid and torrid, but we still talk of temperate climates. It was a simple enough scheme, but a bit general, so in succeeding centuries others added more detail. This led them to attempt definitions of the penguin climate, tundra climate, savannah climate, and many more. We still use some of those names, but others, including penguin climate, have fallen out of favour.

Nowadays scientists have access to reliable data on all aspects of climates and are able to categorize them in many complex ways, but for most of history the only data available referred to precipitation, temperature, and vegetation, so that is what the classifiers used. There were, and are, many schemes for classifying climates. The German climatologist and meteorologist Wladimir Peter Köppen devised the system most often published in atlases and used by geographers, and it is the one most useful to a gardener contemplating some serious relocation. Before you make a final commitment, it might be best to study the Köppen climate map to know what to expect. If you live in a Cfb part of the world and are planning a move to a BWk location, you'll find the growing conditions markedly different.

Köppen first published his classification in 1884, plotting the different climates on an imaginary continent he called *Köppen'sche Rübe* (Köppen's beet). He went on revising it until his final version appeared in 1936, in his later years collaborating with Rudolf Geiger, who continued the work after Köppen's death, so today it is often called the Köppen-Geiger classification.

Köppen began by grouping climates according to temperature. He noted that trees will not grow where the average summer temperature is lower than 10°C (50°F), and there are certain plants that will not survive winter temperatures lower than 18°C (64°F). If the mean annual temperature is –3°C (27°F) or lower there will be frost and probably at least some snow. He used capital letters to designate the climates he derived from this, plus one (B) defined by its aridity.

A	Tropical rainy climates in which the temperature in the coldest month does not fall below 18°C.
B	Dry climates.
C	Warm, temperate, rainy climates in which the temperature in the warmest month is higher than 10°C and in the coldest month is between –3°C and 18°C.
D	Cold boreal forest climates in which temperatures in the coldest month are lower than –3°C and in the warmest month are higher than 10°C.
E	Tundra climates in which temperatures in the warmest month are between 0°C and 10°C.

He then used additional letters to expand on these categories by taking account of precipitation. This produced 29 broad climate types.

Af	Hot and rainy throughout the year
Am	Hot and very rainy in one season
Aw	Hot and dry in winter
BSh	Hot, semi-arid
BSk	Cool or cold and semi-arid
BWh	Hot desert
BWk	Cool or cold desert
Cfa	Mild in winter, hot in summer, moist throughout the year
Cfb	Mild in winter, warm in summer, moist throughout the year
Cfc	Mild in winter, short, cool summer, moist throughout the year
Cwa	Mild in winter, hot in summer, dry in winter
Cwb	Mild in winter, short, warm summer, dry in summer
Cwc	Mild in winter, cool summer, dry winter, rainy summer
Csa	Mild in winter, hot and dry in summer
Csb	Mild in winter, short, warm, dry summer
Dfa	Very cold in winter, long, hot summer, moist throughout the year
Dfb	Very cold in winter, short, warm summer, moist throughout the year
Dfc	Very cold in winter, short, cool summer, moist throughout the year
Dfd	Very cold in winter, short summer, moist throughout the year
Dwa	Very cold, dry winter, long, hot summer
Dwb	Very cold, dry winter, cool summer
Dwc	Very cold, dry winter, short, cool summer
Dwd	Very cold winter, short, moist summer
Dsa	Very cold winter, dry, hot summer
Dsb	Very cold winter, dry, warm summer, high elevation
Dsc	Very cold winter, dry, warm summer, highest elevation
Dsd	Extremely cold winter, winter wetter than summer
ET	Polar climate with short summer
EF	Perpetual frost and snow

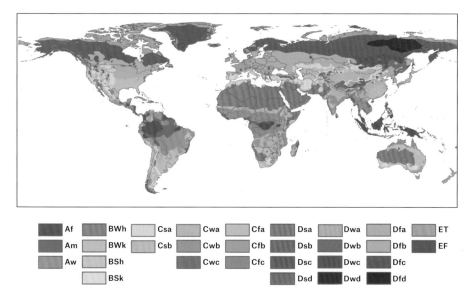

Af	BWh	Csa	Cwa	Cfa	Dsa	Dwa	Dfa	ET
Am	BWk	Csb	Cwb	Cfb	Dsb	Dwb	Dfb	EF
Aw	BSh		Cwc	Cfc	Dsc	Dwc	Dfc	
	BSk				Dsd	Dwd	Dfd	

The Köppen climate classification. Map courtesy of M. C. Peel, B. L. Finlayson, and T. A. McMahon, University of Melbourne.

The map above shows the distribution of these basic climate types. The classification added further detail, using 24 letters to qualify the basic categories. For example, *a* means the temperature in the warmest month is about 22°C (72°F), *f* means there is sufficient precipitation for healthy plant growth in all seasons, *n* means fog is frequent, and *u* means the coldest month follows the winter solstice.

The world's climates

Climates are determined by two principal factors, latitude and proximity to the ocean. To the south of the arctic tundra, where plant cultivation is impractical, there lies a belt of mainly coniferous forest that extends across North America and Eurasia. This is called the boreal forest, after Boreas, the Greek god of the north wind, and in Russia it is known as the taiga. It occurs between latitudes 65° N and 55° N. Summers are short, lasting one to three months, with temperatures that average about 15°C (59°F) but sometimes reach the high 30s (80s). For the whole of the winter, which lasts six to nine months, the temperature remains below freezing. The climate is generally fairly dry. The name *boreal* is appropriate, because this type of forest does not occur in the Southern Hemisphere, where there is very little land in the

The branches of coniferous trees bend under the weight of snow, allowing them to shed the surplus. This prevents the branches from breaking and also exposes the leaves, so they can photosynthesize whenever the temperature rises.

relevant latitude. Forestry is the only commercially significant form of cultivation in the boreal climate.

In temperate latitudes, between the low-latitude edge of the boreal forest and the tropics, the climate varies greatly. The climate of anywhere on or close to the western coasts of a continent will be of a maritime, or oceanic, type. The weather will be very changeable because movements of the polar front will mean that sometimes the air is tropical and sometimes it is polar, but in general the annual temperature range will be smaller than the range found farther inland. Storms will be frequent, especially in winter. Skies will often be cloudy and precipitation will be distributed fairly evenly through the year, but with rather more in winter than in summer. If summers are warm this is a Cfb climate in the Köppen classification, and if summers are short and cool it is Cfc.

Places on the eastern coast of a continent have somewhat drier climates with a greater temperature range. This is because, despite their proximity to the ocean, weather systems approach them from the west, across the continent, so they bring dry air. As each system reaches the coast, however, the winds circulating around its anticyclones and cyclones will blow from the sea for part of the time, bringing air that carries more moisture than the continental air to the west. Such climates are of the Cwa and Cwb type.

Mediterranean climates, in lower latitudes, are named for the lands bordering the Mediterranean, but they occur in all continents in both hemispheres between latitudes 30° and 45°. Winters are rainy and summers hot and dry, so the climate is intermediate in type between that of a maritime west coast and a west coast desert. Droughts are common in summer when maritime tropical (Mt) air masses predominate. Maritime polar (Mp) air masses predominate in winter. These produce frontal systems delivering abundant rain and triggering storms. The temperature range is moderate. These are of the Csa climate type if the summer is hot and Csb if it is warm.

Grassland climates are either temperate or subtropical, and are designated BSh or BSk, respectively, in the Köppen scheme, but shading into BWk, cool desert climates, in the Eurasian steppe. Saskatoon, Saskatchewan, in the North American prairies, has an annual temperature range of 38°C (68°F) and an average annual precipitation of 370 mm (14.6 in.) falling every month, but with summers wetter than winters. The climate is semi-arid with cool summers and cold winters. Ulan Bator (Ulaanbaatar), the capital of Mongolia, has a BSk approaching BWk (cool desert) climate, typical of the interior of the Eurasian steppe grasslands. The annual temperature range is 41°C (74°F) and the average annual precipitation is 210 mm (8.3 in.), falling mainly in summer. Kazalinsk, Kazakhstan, has a temperature range of 41°C and an average annual precipitation of 125 mm (4.9 in.), with dry summers. In both areas the summers are cool, the winters cold.

The African savannah is typical of tropical grasslands. It is of the Aw type, indicating that summers are hot and winters are dry. The temperature range is between 20°C (36°F) and 30°C (54°F) and the annual rainfall averages 250–750 mm (9.8–29.5 in.).

Island climates vary considerably, but it is the surrounding ocean that determines their characteristics. They are moister than any large land area in the same latitude and have a smaller temperature range.

Mountain or highland climates are cooler than those of the adjacent lowlands owing to their elevation. The annual temperature range is lower than that of the lowlands, but the daily range is greater. The highlands are moister up to about 4 kilometres, but drier at higher levels because of the low temperature, cold air holding less water vapour than does warm air. The characteristics of highland climates are determined by elevation rather than latitude, and aspect—the direction the ground faces—controls the amount of direct sunshine and therefore the temperature. Mountain climates occur in all continents, but are always local. They are sometimes designated *H* in the Köppen classification, although this category was not introduced either by Köppen or his colleague Geiger.

Changing climates

We hear a great deal nowadays about climate change. At times it can seem as though the world's climates are ordinarily static, the same century after century, and that the change we have seen during the twentieth century is somehow aberrant and to be feared. But climates have always changed. It is periods without change that are

The remains of stone walls mark the boundaries of abandoned fields on the high ground of Bodmin Moor, in Cornwall. This ground was once cultivated, at a time when the climate was more benign and the demand for farmland stronger.

unusual. At various times in the past, conditions have been warmer than they are today, and cooler, and wetter, and drier.

Historians of climate think in centuries and millennia, and to a geologist a million years is but the briefest instant. All around us there are reminders of past climates. The summer of 2003 was unusually hot and a patch of ice melted in the Schnidejoch Pass, in the Swiss Alps. A couple of hikers walking through the pass that year came across an archer's quiver made from birch bark. Radiocarbon dating revealed the quiver to be 4700 years old. Other items have also been found in that pass, including a late Neolithic leather shoe complete with its laces. In 2010 and 2011 the temperature fell once more and any undiscovered relics are now buried beneath a deep layer of snow. These discoveries prove that the pass was open to travellers at least four times during the last 5000 years. Those were times when the climate in central Europe was warmer than it is today. Swiss glaciers have retreated during the twentieth century, and in doing so they too have revealed items lost in earlier times from the

Early Bronze Age, Iron Age, Roman period, and Middle Ages, as well as human remains from the nineteenth century. On the moors of southwest England there are outlines of fields that were cultivated centuries ago, during times of "land hunger" but also times when the climate allowed cultivation, which is no longer the case. At higher levels there are huge boulders that rolled to their present positions when the frozen ground beneath them thawed and the soil turned to mud. The ground does not freeze so firmly now. There are places such as Upper Teesdale in County Durham, England, where plant communities survive from the most recent ice age.

It is time, therefore, to place the recent climatic change in a historical context, beginning with a large time span, extending back 65 million years, then reducing the span in steps, to the recent ice ages and eventually to the last century. Once up to date, it is appropriate to consider the implications of changes in climate and the feasibility of taking control not of the climate, but of the weather.

How climate changes

Earth's history began approximately 4.6 billion years ago, in an unimaginably distant past, as the planet finally took shape, assembled from countless rocks, dust particles, and gases orbiting the infant Sun. Geologists, whose business it is to reconstruct the history of our planet, divide that long story into a hierarchy of discrete episodes, in descending order called eons, eras, periods, and epochs, most of which have geologically identifiable beginnings and endings. Today we live in the Holocene epoch of the Pleistogene period of the Cenozoic era of the Phanerozoic eon.

The Cenozoic (sometimes spelled Cainozoic) began 65.5 million years ago. That is when a series of changes that altered climates and environments culminated in the arrival of an asteroid, an immense rock about 10 kilometres across, which slammed into the sea off what is now the Yucatán Peninsula of Mexico, releasing a huge amount of energy. The impact wrought changes that caused the extinction of many animal groups, including the dinosaurs—apart from birds, which are direct descendants of dinosaurs, so technically they are also dinosaurs.

During the Cretaceous period preceding the Cenozoic, the dinosaurs had lived in a warm world. In every part of the Earth the climate was similar to the present climate of the tropics. Rainfall was abundant and plants flourished, though most were very different from modern plants, of course. These benign conditions continued during the first epoch of the Cenozoic, the Palaeocene

(65.5–55.8 million years ago), when palm trees grew in Greenland and there were mangrove forests at 65° S on the edge of the Antarctic Circle, but changes were happening. Sea levels were falling, exposing dry land in North America, Africa, and Australia, and isolating South America from Antarctica.

The following epoch, the Eocene (55.8–33.9 million years ago), began with the most dramatic episode of global warming ever recorded. It is called the Palaeocene–Eocene Thermal Maximum and in the space of 100,000 years the temperature rose by 5–7°C (9–13°F) and remained at that level for the next 200,000 years, after which the world grew slowly but steadily cooler, although with several warmer intervals. As the global climate grew colder, ice began to appear at the poles and the Antarctic ice sheet began to expand rapidly at the end of the Eocene.

During the Oligocene epoch (33.9–23.03 million years ago) the continents were moving and their movements altered ocean currents. Water flowed southward from the Arctic Ocean, and the separation of Antarctica from South America allowed the circumpolar current to flow. This isolated Antarctica and it is the continent's isolation that caused the ice sheet to expand and that helps maintain it to the present day. The expansion of the ice sheet led to a fall in sea level, exposing more dry land. Climates generally were cooler and drier than they had been, and more seasonal. About 25 million years ago the first grasses appeared, well suited to the cooler, drier conditions, and as they spread across the continents during the next few million years, herds of grazing mammals became larger, and the grasses provided shelter and nesting material for small mammals and birds.

Then global temperatures began to rise once more in the early part of the Miocene epoch (23.03–5.3 million years ago), and tropical forests expanded, but the warming was short-lived. Temperatures fell again as the Antarctic freeze intensified. Mountain chains grew—the Cascades in western North America, the Andes in South America, and the Himalayas in Asia. These altered the flow of air, and climates became drier. Scrub plants, able to cope with the increasing aridity, made their first appearance. Sea levels continued to fall. Inland seas dried out and dry land linked Africa, Eurasia, and North America.

The Pliocene epoch (5.3–2.588 million years ago) followed the Miocene, and the world's climates continued to become cooler and drier. Tundra vegetation appeared in high latitudes, with boreal forest and grasslands where the climate was warmer. As sea levels continued to fall, dry land was exposed, connecting North and South America. This allowed animals to migrate between the continents and it also separated the marine organisms of the Atlantic and Pacific.

A periglacial landscape, where repeated frost action has arranged
stones into patterns.

By two million years ago ice covered both poles. The world was then primed for
the series of ice ages that were about to commence.

Ice ages and interglacials

Lying scattered in Swiss mountain valleys there are large boulders made from
granite and quite different from the underlying bedrock that are marked with
long, parallel scratches. For many years these boulders puzzled a number
of scientists familiar with that region of Europe. They were much too heavy
to have been carried by people—and for what conceivable purpose? Could a
flood have transported them? That seemed unlikely. Jean-Pierre Perraudin,
a hunter who knew the mountains well, had an idea that glaciers might have
moved them, but in the early nineteenth century no one imagined that glaciers could move. But Perraudin persisted. He consulted geologists and one or
two took his idea seriously. Eventually they persuaded Louis Agassiz, already
famous for his studies of fossil fishes and professor of natural history at the
University of Neuchâtel, to consider it.

In 1836 and 1837 Agassiz spent his summer holiday with friends in a hut
they built on the Aar Glacier, studying the rocks on either side of the ice.
They also drove a straight line of spikes into the ice across the glacier. They
found parallel lines of scratches (striations) in the ice bordering the glacier,
and when Agassiz returned in 1841 he found the line of stakes was no longer
straight. The centre of the line bulged in the direction of the foot of the glacier.

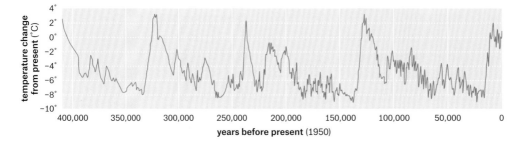

Ice cores from the Russian Vostok Station in Antarctica record changes in average temperatures over the last 400,000 years. These record a series of ice ages and interglacials as departures from the present mean temperature. To convert the graph values to actual temperatures, add 14°C.

Clearly, the ice had moved, and, Agassiz reasoned, friction with the rocks at the sides slowed the movement at the edges, but it was that movement which made the striations. Agassiz was convinced that glaciers move, albeit slowly, and he pursued his study of the phenomenon, finally coming to believe that a thick layer of ice had once covered all the land from the North Pole to the Mediterranean. He believed there had been what he called a Great Ice Age, and he toured Europe and the United States lecturing on the subject and searching for more evidence.

The idea was revolutionary and Agassiz was almost right. He was mistaken in thinking that the ice had reached as far as the Mediterranean, and we now know that his Great Ice Age was a series of ice ages interrupted by warmer interglacial periods. These began toward the end of the Pliocene epoch and continued throughout the Pleistocene (1.806–0.1143 million years ago). The end of the most recent ice age marks the transition from the Pleistocene to the Holocene epoch in which we live today, although the implication that the ice ages have ended is misleading. Climatically we are still living in interglacial in what is still an age of ice ages, and one day the ice will return.

Much of our knowledge of changing temperatures comes from studies of cores of ice drilled from the Antarctic and Greenland ice sheets. These ice cores contain dust particles, tiny gas bubbles, and variations in the ratios of the two principal isotopes of oxygen (^{16}O and ^{18}O) from which scientists can reconstruct a history of worldwide temperature. Cores drilled at the Russian Vostok Station, Antarctica, provide a record going back 400,000 years. The graph above summarizes that record as departures from the present global average temperature of 14°C (57°F), positive values indicating temperatures warmer than those of today and negative values cooler temperatures. As the graph shows, there have been many oscillations between warm and cool, ice ages and interglacial periods. It also shows that the present interglacial is cooler than the four that preceded it.

Although the ice ages were global in extent, there were differences in timing and intensity from place to place and the evidence for them was discovered at various locations. Consequently, although the ice ages are more or less contemporaneous, they have different names in different parts of the world. The table below lists them with their North American, British, and northwest European names and approximate dates.

PLEISTOCENE ICE AGES AND INTERGLACIAL PERIODS

Approximate date (thousands of years ago)	North America	Britain	NW Europe
10–present	*Holocene*	*Holocene*	*Holocene*
75–10	Wisconsinian	Devensian	Weichselian
120–25	*Sangamonian*	*Ipswichian*	*Eemian*
170–120	Illinoian	Wolstonian	Saalian
230–170	*Yarmouthian*	*Hoxnian*	*Holsteinian*
480–230	Kansan	Anglian	Elsterian
600–480	*Aftonian*	*Cromerian*	*Cromerian complex*
800–600	Nebraskan	Beestonian	Bavel complex
740–800		*Pastonian*	
900–800		Pre-Pastonian	Menapian
1000–900		*Bramertonian*	*Waalian*
1800–1000		Baventian	Eburonian
1800		*Antian*	*Tiglian*
1900		Thurnian	
2000		*Ludhamian*	
2300		Pre-Ludhamian	Pre-Tiglian

Italic names indicate interglacial periods, Roman names indicate ice ages. Dates are increasingly uncertain for the earlier periods and no evidence has been found for ice ages in North America prior to the Nebraskan. The only evidence for the Thurnian ice age and Ludhamian interglacial comes from a single borehole at Ludham, Norfolk, England.

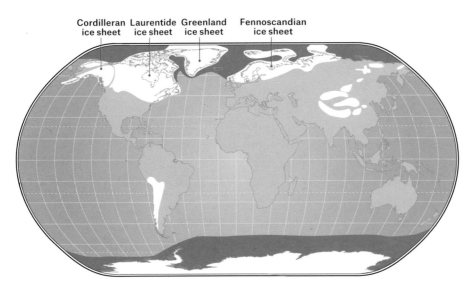

At their maximum extent, ice sheets kilometres thick covered large areas of the Northern Hemisphere and parts of the Southern Hemisphere.

During the most recent ice age, the Devensian (Wisconsinian), the Laurentide and Cordilleran ice sheets, kilometres thick, covered most of North America as far south as the southern edge of the Great Lakes and the Fennoscandian ice sheet covered northern Eurasia including all of northern Britain. An ice sheet covered the southern Andes in South America and there were smaller ice sheets in the Himalayas, southeastern Australia, and New Zealand. Ice sheets lie on land and, as the map above shows, beyond the edges of the northern ice sheets the sea was permanently frozen.

If you fear the onset of a new ice age, never mind about a cold winter, or even a succession of cold winters. Ice ages begin in summer, not winter, so watch out for a steady decrease in summer temperatures. The first sign to cause you some alarm will be the persistence of snow throughout the summer. At first it will be only small patches of snow here and there, on hillsides perhaps, and, of course in high latitudes. On mountains that retain a snowcap all year, the cap may be just a bit bigger than it was last year. Snow has a high albedo and by reflecting sunshine it chills the air above it and the adjacent ground. The following summer just a little more snow survives than survived the previous summer. Year by year, and quite gradually, the area of permanent snow expands and each winter more snow falls on top of last winter's snow, so the layer of snow grows thicker as well as wider.

Glacial ice does not form in the same way as the ice on a winter pond or in a freezer, but more in the way fresh snow turns to ice after cars have driven over it. It begins as snow, which falls as individual flakes that pack together loosely, with many air spaces in the structure. As the layer of snow grows thicker its weight compresses the lower layers, squeezing out most of the entrapped air. The snow becomes denser and with sufficient time and pressure it turns into ice.

One day the ice will return and now you know what to look out for. But fear not, for when scientists say a new ice age is overdue and imminent, they mean we should expect it any time in the next few thousand years. They're not suggesting it will happen next week.

Minoan, Roman, Medieval, and current warm periods

In the first century CE the Roman Emperor Domitian banned the production of wine north of the Alps, presumably to protect vine growers in southern Italy, but in about 280 the Emperor Probus rescinded the ban. During the first centuries of their occupation, Romans based in Britain had relied on imported wine, but imports appear to have ceased in about 300 and the Romans are known to have introduced viticulture to Britain. There were productive vineyards in northern England and it is possible that Britain had become self-sufficient—and a serious competitor to the Italian producers. The change was probably due to changing economic and commercial conditions, for it must have been difficult and costly to transport wine over such a distance, but at least partly it was due to the climate becoming warmer. Between about 250 BCE and 400 CE, known as the Roman Warm Period, temperatures throughout the Northern Hemisphere and probably globally were very similar to those of today and at least in some places a little warmer. North Africa and parts of the Near East were also moister. North African farmers grew the grain that fed Rome, and the city of Petra, in Jordan, flourished between 300 BCE and 100 CE in what has since become desert.

In the Middle Ages, when the kings of England also ruled lands in France, French wine producers tried to persuade their rulers to agree a treaty forbidding the import to France of English wine. Not only was there a flourishing wine industry in England, with vineyards as far north as Lincolnshire, but clearly the wine was of good enough quality to rival French vintages. Some of the English vineyards were sited in what are now frost hollows, indicating that it was rare for late frosts to damage the blossoms.

The Middle Ages was a time when temperatures were slightly higher than those of the late twentieth century. It is known as the Medieval Warm Period and lasted from about 950 until 1250. It was when Norse seafarers travelled widely, even as far as North America, and Norse people established settlements there and also in Greenland, where they raised cattle, sheep, and goats, and managed to grow small amounts of barley, which they would have needed for brewing beer. Forests of birch bordered the fjords and willow and grass grew on the hills. The settlers combined farming with fishing for cod, which was abundant. Ruins have been found in Greenland of more than 300 farms, 22 churches, and a convent dating from that time.

The Norse account of these settlements in the *Landnámabók*, written in about 1125, tells of Thorkel Farserk who planned a feast for his cousin Erik the Red. He needed mutton, but all the sheep were on an island and there was no serviceable boat to hand. So Thorkel swam more than three kilometres to the island, grabbed a full-grown sheep, and swam back with it. Scientists who have studied the physiological effects on long-distance swimmers reckon that this feat would have been impossible in water cooler than about 10°C (50°F). Today the temperature in that stretch of water averages 3–6°C (37–43°F). We can also be sure the weather was warmer then because Norse Greenlanders buried their dead in deep graves dug in ground that is now permanently frozen and hard as iron.

While the Greenlanders were raising sheep, their cousins in Iceland were also growing barley and oats. In Norway, farmers were growing cereal crops, most likely barley, at 69.5° N, inside the Arctic Circle, and clearing hillside forests to make farmland up to 200 metres higher than they had been able to cultivate earlier. In Britain, too, farmers were cultivating land at higher elevations.

The Medieval Warm Period was just one of a series of similar episodes that began soon after the end of the most recent ice age with a period known as Holocene Warming a. It lasted from about 11,600 to 8500 years ago, and the Holocene Warming b period lasted from about 8000 to 5600 year ago.

The earliest European civilization arose in Crete. It traded with Egypt and Mesopotamia, developed writing, and was the forerunner of the Greek civilization. It was named Minoan after its legendary king Minos and it lasted from about 3650 BCE until 1100 BCE. For part of that long period, from about 3500 BCE to 3200 BCE, the Minoans lived through another warm period, the Minoan Warm Period.

As the graph opposite shows, the Minoan, Roman, and Medieval Warm Periods were all warmer than the Current Warm Period, at least in Greenland. The graph is based on data acquired from ice cores drilled as part of the Greenland Ice Sheet Project (GISP).

Historically, warm periods have been times of prosperity and advances in civilization. The change in temperature is slight, a matter of no more than a

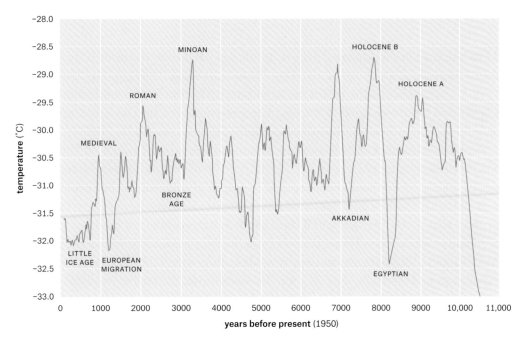

Changing temperatures over the last 11,000 years, based on data from ice cores drilled in central Greenland as part of the Greenland Ice Sheet Project 2 (GISP 2).

degree or two, but it extends the growing season. Harvests are larger and more dependable, people are healthier because they are better fed, and resources can be released for major projects. The Minoans could sail their world, exporting their surpluses and bringing home goods they could not produce themselves, because the sailors and shipbuilders did not have to spend their days struggling to grow food. The Romans expanded their empire. And in the Middle Ages Europeans built the most magnificent cathedrals the world has ever seen.

European migration period and the Little Ice Age

Warm periods are separated by cooler times and just as warm temperatures bring prosperity, cool periods bring hardship. The troubles associated with the more recent cool periods are well documented and studies of cores drilled from ice sheets have revealed earlier ones, before history came to be written. These are shown in the graph, the earliest being the Egyptian Cool Period, also known as the 8.2 kiloyear event. This followed Holocene Warming a, and it lasted from about 8500 to 8000 years ago. The cooling affected the entire world, with temperatures falling by 1–5°C (2–9°F).

No one can tell what suffering that cooling may have caused, but the

Akkadian Cool Period changed history. It followed Holocene Warming b, and lasted from about 5600 to 3500 years ago. Despite its name, the Egyptian Cool Period occurred before the rise of the Egyptian civilization, but the Akkadian Cool Period saw the collapse of the Egyptian government of the Old Kingdom as failures of the annual Nile flood brought famine that led to social breakdown.

Glaciers advanced and there were prolonged droughts throughout the Northern Hemisphere. Harvests failed, triggering unrest as farmers abandoned their dust-dry land and wandered in search of somewhere they could make a living. Soldiers deserted, weakening armies and allowing outsiders to invade. That is the fate that befell the Akkadian Empire centred on the city of Akkad in what is now Iraq. Invaders from the Zagros Mountains captured Akkad about 4130 years ago and destroyed it so thoroughly that modern historians cannot be certain of its location. Three centuries passed before people once more settled on the northern plains of the vanished empire.

The Minoans made bronze tools and fought with bronze weapons. They were people of the Bronze Age and the change in climate that coincided with the collapse of their civilization was the beginning of the Bronze Age Cooling, a period lasting from 3200–2500 years ago. The collapse of Bronze Age cultures was widespread. Glaciers advanced in Europe and North America and in Central and South America temperatures fell sharply around 3500 years ago. During the preceding warm period, settlements had developed around lakes in central Europe with buildings raised on piles. As the climate deteriorated, catastrophic floods destroyed these settlements and may have been responsible for driving Phrygian people from the Hungarian plain into Anatolia, where they contributed to the collapse of the Hittite Empire. Other migrants moved into Italy, their descendants joining the ancestors of the later Etruscans and Romans. In the Near and Middle East, seafaring raiders, known as the Sea People, attacked Hittite, Assyrian, Egyptian, and other coastal settlements. The Egyptian Empire came under attack and finally collapsed around the commencement of the Bronze Age Cooling, during the reign of Ramesses VI (1145–1137 BCE). Most famously of all because the story is told in Homer's *Iliad* and several Greek tragedies, it was during this period that the city-state of Troy collapsed and its fabled walls, nine metres high, fell.

The Bronze Age Cooling brought catastrophe on the scale of the collapse of the Western Roman Empire. The Roman collapse followed a long period of decline and when, finally, Rome lost all political and economic control of its former territories, Europe entered what used to be called the Dark Ages. Most historians no longer use that derogatory term, preferring to call it the

A roofless house is all that remains of an abandoned Norse settlement in Greenland where a farming community made a living during the Medieval Warm Period.

European Migration Period, because it was a time when there were large movements of population.

Why were people moving? The withdrawal of the Roman legions meant there were no longer troops to protect farms and settlements from marauding tribes for whom there were rich pickings to be had, but at least in part the migrations occurred because from about 500 to 900 CE temperatures fell sharply, the glaciers advanced, and the crops failed repeatedly. Central European and Scandinavian forests expanded on to abandoned farmland. The Black Sea froze in 800, and in 829 the River Nile froze. The Rhine may have frozen from time to time, allowing migrants to cross freely.

The most recent cooling began in about 1300, ended in about 1850, and was the coldest episode since the Younger Dryas, a bitterly cold partial return of the ice age 12,900–11,600 years ago. The Norse Greenland settlements failed and most of the settlers died when the sea froze, preventing fishing and trade with Norway. In several years ice blocked the Denmark Strait, between Iceland and Greenland, in summer. Glaciers overran Icelandic farms and volcanic

The Thames froze many times during the Little Ice Age and Londoners walked, skated, played games, and, of course, set up stalls on the ice to sell their wares in frost fairs like this one.

eruptions beneath the glaciers released water that flooded large areas and, as the water retreated, left them buried in river-borne sand and gravel. People turned to fishing, but the cod also failed them. It was a period of sea storms more violent and more frequent than those of today. In the autumn of 1697 a great gale in North Uist, in the Hebrides, buried the settlement of Udal in sand. Udal had been inhabited for almost 4000 years. Between 1690 and 1728 Eskimos in kayaks were seen several times in the Orkney Islands, off the north coast of Scotland, and once in Aberdeen. The Thames froze several times. In 1536 Henry VIII travelled along the river by sleigh and the first frost fair was held on the ice in 1608. The last was in 1814, when an elephant walked across the river.

But colder weather also meant failed harvests, which meant famine. The Great Famine affected northwest Europe from 1315 to 1317. Malnutrition reduced resistance to disease and the Little Ice Age was a time of terrible disease. The worst was the Great Pestilence, later called the Black Death, an outbreak of bubonic, pneumonic, and septicaemic plague that entered Sicily in 1347 and spread throughout Europe, killing up to 200 million people before it died out in 1351. It also spread eastward across the Middle East.

The Little Ice Age was also a time of war. The Hundred Years War, from 1336 to 1453, in which England and France fought for dominion over Flanders, had complex causes, but it was one more of the tragedies that blighted the fourteenth century.

CHANGING CLIMATES

The message of history is clear. Periods of warming are associated with peace, prosperity, and cultural advances. Cold periods bring famine, disease, and war.

Climate since 1900

The Little Ice Age came to an end in the latter part of the nineteenth century, but the recovery from it began much earlier. Temperatures began to rise soon after 1700, albeit hesitantly, and in the 1730s Europe enjoyed temperatures similar to those of the twentieth century. But the improvement was not sustained. In the winter of 1708–9 people walked across the frozen Baltic and in January 1716 the Thames froze so hard that a high spring tide raised the ice by 4 metres, complete with its frost fair.

A more sustained rise in temperature began toward the end of the nineteenth century. Twentieth-century changes in temperature are usually measured from about 1850, which is when reliable records first became available, and more recent changes are measured as anomalies, which are deviations from a climate normal such as 1979–2009. Obviously, the quality of measurements has continued to improve since 1850, but calculating the average temperature over the whole world is far from simple. Significant changes in temperature are counted in tenths of a degree and must allow for local effects that distort them, such as the warmth generated in even quite small urban areas, and scientists must ensure that observers everywhere are using compatible instruments to take readings at the same times every day. Weather stations record observations appropriate for weather forecasting, and forecasters are interested in local changes over a matter of days, whereas climate scientists need information about extremely small changes that occur over a very large area over decades. Yet climatologists must do the best they can with data from ordinary weather stations. Instruments carried aloft by weather balloons do better. These are launched at midnight and noon UT (Universal Time) everywhere in the world. They avoid the problems of siting that can affect surface stations, but large areas of the world are not covered. The most reliable series of readings began in 1979 with the launch of the first weather satellite carrying an instrument called a microwave sounding unit. Polar-orbiting satellites now carry the successors of these instruments and together they measure the middle tropospheric air temperature over the whole world several times each day.

Since 1850 the global average temperature has risen by 0.7°C (1.3°F) but, as the graph on page 86 (above) shows, the rise has not been a steady one. At first the temperature fell, but then rose from about 1910 until it reached a peak in

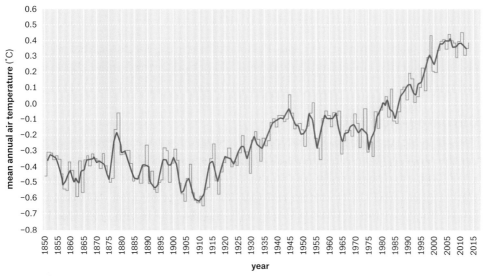

Global temperatures are measured as variations from the average temperature between 1979 and 1988. The thick line is a three-year running average of the annual temperatures shown by the thin line. Data from the Hadley Centre (UK Met Office) and the Climatic Research Unit, University of East Anglia (HadCRUT).

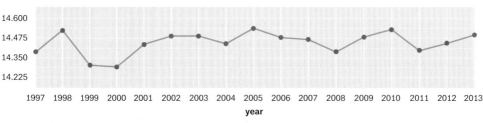

Since the beginning of 1997, the global mean atmospheric temperature has risen and fallen a little from year to year, but overall it has not changed. The world is no warmer in early 2014 than it was in early 1997. Data from the Hadley Centre (UK Met Office) and the Climatic Research Unit, University of East Anglia (HadCRUT).

1945. Then the world cooled until the warming resumed in 1976. Part of this warming was the continuing recovery from the Little Ice Age, but the latter part, from 1976, is usually attributed to the steady rise in the atmospheric concentration of carbon dioxide. Carbon dioxide is a greenhouse gas—it absorbs infrared radiation—and it has been entering the air through the burning of fossil fuels because it is a by-product of combustion of carbon-based fuels, the kilning of limestone to make lime used in the manufacture of cement ($CaCO_3$ → $CaO + CO_2$), and changes in land use that release carbon dioxide from the soil. There is no doubt that the amount of carbon dioxide in the air has been increasing or that carbon dioxide absorbs radiation, and it follows that the increase will have raised the temperature of the lower atmosphere. But there are two problems.

The first concerns the changes in temperature between 1930 and 1976. The 1930s were the years of the world depression, when factories were closing and industrial production was in decline, and with it the rate of emission of carbon dioxide. Yet that was a time when temperatures were rising strongly. The period between 1945 and 1975 was when the world was repairing the damage inflicted during the war. It was a time of full employment and rapid economic expansion, with power generation and manufacturing powered by burning coal and oil. Carbon dioxide levels rose dramatically throughout the war, when temperatures were continuing to rise, and emissions went on rising during the economic boom that followed the war, yet the boom saw global temperatures fall.

The second problem arises from the fact that the global average temperature did not rise between January 1997 and October 2014. As the second graph (opposite, below) shows, at the beginning of 1997 the global mean temperature was about 14.4°C (57.9°F) and at the end of 2013 it was about 14.5°C (58.1°F), but over these 17 years it varied very little and always within the margins of error of instrument readings and data processing, apart from a temperature spike in 1998 caused by an extremely strong El Niño event. Neither have temperatures fallen, however. Global warming has certainly occurred and we are now in what is called the Current Warm Period, comparable to the warm periods that preceded it. So when people say, as they often do, that this or that spell of bad weather is a sign of global warming, they should point out that in recent years there has been no warming.

Temperature and extreme weather

Predictions of what the world's climates may be like a century or so from now are based entirely on millions of highly complex mathematical calculations—mathematical models—performed by supercomputers. The scientists constructing the models begin with a three-dimensional grid covering the planet at the surface and at several levels above it. They enter values for atmospheric conditions at each grid intersection, introduce a change, for example in the chemical composition of the air, then calculate the effect this will have at all the grid intersections. The model then progresses the way the intersections influence their neighbours. With each change in conditions, all the calculations must be repeated and so, step by step, a picture emerges of the way the climate will evolve. Inevitably, however, the models have deficiencies. Because of the scale of the grid they must make assumptions about events, such as the formation of clouds, which take place on a smaller scale. There are areas of atmospheric science that are not well understood. The modellers must use estimates for all of these and because the development of weather systems is acutely sensitive to small variations in starting conditions, the models should be regarded as valuable research tools, but poor predictors.

The models predict that steadily rising concentrations of greenhouse gases will cause a sustained rise in atmospheric temperature, which implies an increase in the amount of energy held in the atmosphere. If the atmosphere has more energy, the models predict that its behaviour will become more violent. There will be more extreme weather events, such as storms, and they will be more intense.

An obvious way to check the reliability of such predictions is to see what happened in the recent past and also during prolonged periods that were warmer or cooler than those of today. In 2009, R. Allen and colleagues published the results of their study of records of autumn and winter storms over the British Isles, decade by decade between the 1920s and 1990s. They found there were more autumn storms in the 1920s than in the 1990s, but more winter storms in the 1990s than in the 1920s. Overall, however, there were rather more storms in the 1920s and fewer in intervening decades, but with a rise beginning in the 1970s. In 2004, A. Dawson and colleagues published their findings after examining records for gale-days—days on which gales were recorded—from five sites in Scotland, Ireland, and Iceland over the last 120–225 years, and then compared these with records over approximately 2000 years from the same region. They found that gales were more frequent in the nineteenth and early twentieth centuries than they are today, and over the 2000-year period winter storms were more common during cold than warm periods and were at a minimum during the Medieval Warm Period.

Some of the intense storms that battered Great Britain during the winter of 2013–14 coincided with high spring tides, consequently producing large tidal surges as powerful onshore winds drove the water ashore. There are records of tides and sea levels, including exceptionally high levels, and in 2002, P. L. Waterman and colleagues reported their analysis of four discontinuous sets of data from Liverpool waterfront covering the period from 1768 to 1999. They were looking for changes in the annual maximum high water, the tidal surge component of high water, and maximum surges at high water. They found that the maximum surges at high water declined at a rate of 0.11 ± 0.04 metres per

IPCC ESTIMATES OF EXTREME WEATHER LINKED TO CLIMATE CHANGE

Event	Probability
Overall increase in warm days and nights, and decrease in cold days and nights	very likely
Increase in warm spells and heat waves	medium
Increase in tropical cyclones (hurricanes and typhoons)	low
Increase in tornadoes and hailstorms	low
Increase in droughts	medium
Increase in floods	low*
Increase in extreme high water due to sea-level rise	likely

*In the case of floods even the sign was unclear—it was impossible to say whether they were becoming more or less frequent.

century, but there was no statistically significant trend in the other parameters. At Sefton, on the coast of northwest England, there is a 16-kilometre stretch of coastline for which there are comprehensive records back to the 1890s of tide heights, surge heights, wave heights, and wind speeds. L. S. Esteves and colleagues analyzed these records and reported in 2011 that they found no evidence of enhanced storminess or increases in surge heights or extreme water levels, and while there were seasonal variations and changes on a scale of decades, there was no statistically significant trend over the whole period.

There are many similar studies showing that storms are not becoming more frequent or more severe, and that perhaps we should not expect them to. The historical evidence strongly suggests that prolonged cool periods are much stormier than warm periods.

The Intergovernmental Panel on Climate Change (IPCC) is the body established under United Nations auspices in 1988 to examine the evidence for climate change resulting from human activities and the implications of such change. In the *Summary for Policymakers* of their most recent report, they estimate the likelihood of a link between recent extreme weather events as a result of climate change.

So far, then, there is no evidence that storms and other extreme weather events are becoming more frequent. Will they become more frequent in years to come? We have no way of knowing, but it seems that earlier warnings of dramatic increases in storms, floods, and droughts were grossly overstated.

Warmth and a longer growing season

Phenology is the study of the timing of natural events. Its practitioners, the phenologists, their geographic locations precisely known, record the dates each year on which particular plants open their leaf buds, flower, set seed, and drop their leaves. They note the first appearance of frogspawn in ponds and ditches, when the spawn hatches into tadpoles, and when the tadpoles vanish because they have metamorphosed into frogs. And, of course, they never fail to record hearing the first cuckoo in spring.

In practice it is less whimsical. These data are collected by professional gardeners, organized across Europe by the International Phenological Gardens of Europe project, founded in 1957 and coordinated at the the Humboldt University of Berlin. The gardens supplying records are similar in type, most being level with meadows and some trees, and they grow cloned species of shrubs and trees to eliminate hereditary variability. Each garden has a parent garden nearby where plants are propagated and bred for growing in the phenological garden, and there must be an official weather station in the immediate vicinity. There are 89 European gardens located in 19 countries from Macedonia to Scandinavia, and from Macedonia to Portugal.

More general data are also gathered from satellite observations. That is possible because the burst of leaves in spring and leaf senescence in autumn alter the colour of the land in ways that satellite instruments are able to detect. They take photographs, usually in false colours that emphasize the changes. Green areas appear red, for instance, because vegetation reflects strongly in the infrared part of the spectrum.

Phenological observations record changes in the length of the growing season. The growing season is defined as the period that begins on the first of five successive days when the average daily temperature is higher than 5°C (41°F) and ends on the last day prior to a series of five successive days when the average daily temperature is lower than 5°C. The threshold of 5°C is used because most plant growth ceases below this temperature.

In most parts of the world, the growing season became longer during the twentieth century. Britain has the most comprehensive set of temperature data, called the Central England Temperature. This records monthly temperatures since 1659 and daily temperatures since 1772 across a triangular area of England extending from Lancashire to London to Bristol. These records show that the growing season is now about 20 days longer than it was in 1772, and the present growing season, of an average 274 days, is 21 days longer than the 1961–1990 average, although the average has not increased since 2000. The increase is due mainly to the start of spring, which is four or five days earlier than it was between 1980 and 2000.

Similar increases have occurred in most other parts of the world. In Europe as a whole the growing season is 19 days longer than the 1982–2000 average and it is 18 days longer over Eurasia. In North America spring is commencing about 8 days earlier and the growing season is 12 days longer, ranging from about 14 days in the north-central United States to about 7 days in the east-central United States. In the world as a whole the growing season is about 12 days longer than the 1981–1991 average.

Not all growers have benefited, however. On the Kola Peninsula in northwest Russia the growing season in the taiga has actually decreased by 15–20 days. When M. V. Kozlov and N. G. Berlina checked the records from 1930 to 1998 they found that the start of permanent snow cover began 13 days earlier in 1998 than it did in 1930, the snow around tree trunks melted in spring 16 days later, the summer snow-free period decreased by 20 days, and the period when lakes were free of ice decreased by 15 days. Not surprisingly, the two scientists were startled by results that contradicted the warming they had expected to find and at first thought they must have been mistaken. But they checked and double-checked and were forced to conclude that the growing season on the Kola Peninsula really had shortened.

The Kola result is an anomaly, of course. Elsewhere the growing season is significantly longer than it was. A longer growing season means larger harvests and up to three more weeks to enjoy the garden. There is also a second implication. If the longer season is sustained, in time the land may store more carbon than it did, partly counterbalancing the increase in the atmospheric concentration of carbon dioxide.

The lengthening of the growing season is due to the rise in temperature over the course of the twentieth century and it is interesting to note that such a small change, of only 0.7°C (1.3°F), can produce such a large effect. If present average temperatures are maintained then we can expect the growing season to remain at its present length. If the temperature rises or falls, however, then the length of the growing season will follow it.

Carbon dioxide fertilization effect

Life is based on carbon, at least on this planet and, because of carbon's remarkable capacity for forming bonds with other elements, perhaps on other planets as well. Our bodies are made mainly from carbon, as is the food we eat. We obtain that carbon directly or indirectly from plants, and the plants obtain it from carbon dioxide they take from the air and combine with hydrogen to make sugars by the process of photosynthesis. It all begins with carbon dioxide, therefore, and when living organisms release chemically stored energy by oxidizing carbon, they release carbon dioxide into the air, allowing the cycle to continue.

Carbon dioxide is the basic food on which almost all life depends. To describe it as a pollutant is clearly ludicrous and those who do so betray their biological illiteracy. Removing an environmental pollutant is beneficial, but if by some magical process we were able to remove most of the carbon dioxide from the air all life would cease. Happily, removing more than a trace of carbon dioxide is completely impossible. Were we to do so, respiration by the aerobic bacteria consuming the remains of all those dead organisms would return all of the carbon dioxide to the air.

Carbon dioxide comprises approximately 0.04% of the present atmosphere and there have been times in the past when the air contained less than it does today and much more. About 500 million years ago the concentration was about 0.8%. Then it fell until about 200 million years ago, when it rose to 0.16–0.20%, since when it has slowly decreased. The point here is that plants evolved in an atmosphere containing a much greater proportion of carbon dioxide than it contains today. Indeed, it was the expansion of photosynthesis, first in cyanobacteria, then in algae, and finally in vascular plants, that drew down carbon dioxide from the atmosphere in exchange for oxygen. So the ancestors of modern plants had easier access to carbon dioxide than their descendants. In a word, modern plants are deprived of carbon dioxide. Commercial growers are well aware of this and they routinely add carbon dioxide to the atmosphere in their greenhouses to a level of about 0.1%—2.5 times the concentration of the outside air.

That being so, we might expect outdoor plants to grow more vigorously in air enriched in carbon dioxide, and this is the case. It is called the carbon dioxide fertilization effect and it has been studied extensively.

The best way to measure the effect under conditions as natural as possible is through a free-air carbon dioxide enrichment (FACE) experiment in which vent pipes are arranged in a circle around the growing plant. The pipes release carbon dioxide and are controlled by computers that adjust the flow to take account of changes in the wind speed and direction by reducing the discharge on the downwind side and increasing it on the upwind side, so the plant in the centre of the circle is exposed to a constant level of carbon dioxide, so far as that is possible. Similar plants are grown nearby in ambient air as a control.

Hundreds, possibly thousands, of FACE experiments have been performed over about 25 years on a wide range of plants. In most of them the concentration of carbon dioxide was increased by 200 parts per million, which is about 50% above the present background level. On average, these show that with a carbon dioxide concentration of 600 parts per million (0.06% of the atmosphere), leaf photosynthesis increases by about 30% and dry matter production increases by about 17%. In other words, the plants grow much better in air enriched in carbon dioxide. The effect varies from species to species. All plants benefit to some degree but trees benefit most. In plants other than trees, however, there is a decrease of an average 7% in the nitrogen content of the leaves.

Temperature also has an effect. At temperatures below 25°C (77°F) plants absorbed 28% more carbon dioxide from the enriched air, but at temperatures higher than 25°C they absorbed 45% more.

Biologists consider FACE experiments the best way of measuring plant responses to changing levels of carbon dioxide, but this method may underestimate the extent of the response. That is because with even the most sensitive computer control, changes in the wind make the carbon dioxide level fluctuate quite widely over a time scale of less than a minute. Tested experimentally, such fluctuations were found to decrease the uptake of carbon dioxide. It is likely, therefore, that the benefit from increased carbon dioxide levels is greater than the FACE experiments indicate.

There is a further, very important benefit from higher levels of carbon dioxide. Plants exchange gases through pores—stomata—in their leaves. While the stomata are open to allow carbon dioxide and oxygen to enter and leave the plant cells, water evaporates from the leaf tissues and is replaced by water drawn upward from the soil through the plant xylem. This is the process of transpiration. When the air contains a higher concentration of carbon dioxide, the stomata need not remain open for so long in order for the plant to obtain the amount it needs, and with the stomata open for shorter periods there is less time for water to evaporate. FACE experiments have shown that with a 50% increase in carbon dioxide concentration the efficiency with which plants utilize water increases by an average 81%. If higher temperatures due to climate change make soils drier in years to come, enrichment of the carbon

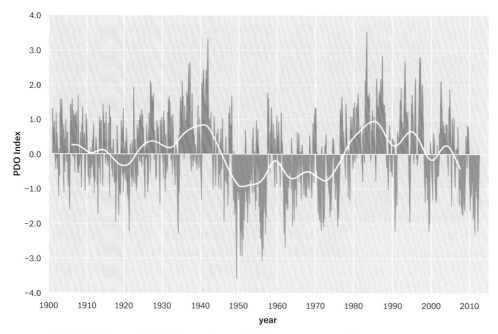

The PDO index, calculated from changes in the temperature of the sea surface in the North Pacific, has gone through a series of positive (red) and negative (blue) periods, shown by the vertical bars representing monthly values. The white line shows the average changes.

dioxide content of the air will allow plants to compensate by using water much more efficiently.

With so much emphasis by environmental campaigners and the media on the grim future they suppose rising levels of carbon dioxide will bring, it is worth pointing out that not all the effects are negative. The evidence for carbon dioxide fertilization is overwhelming.

Greenhouse gases, sunspots, and clouds

Every few years an El Niño event, when the sea-surface temperature changes in the equatorial South Pacific, affects the weather over a large area. El Niño, or more correctly the full El Niño—southern oscillation (ENSO) cycle, may be linked to another cyclical change in the North Pacific. This one occurs approximately every 30 years and it is known as the Pacific Decadal Oscillation (PDO). The PDO involves a shift between two different atmospheric circulation patterns and it manifests as a change in the sea-surface temperature. Scientists monitor the temperature, remove short-term factors to reveal the underlying

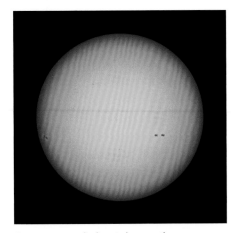

Sunspots are dark patches on the surface of the Sun where the temperature is lower than in the surrounding area. Their number increases and decreases in a cycle of about 11 years.

trend, and report the changes as an index, that is, as departures from a baseline value. The graph opposite shows the monthly PDO index values from 1900 to 2013, with a line showing the averaged trend. Positive departures, in red, indicate that the sea is warm, and negative departures, in blue, indicate cool periods.

The change in temperature is small, but the cycle is interesting. The sea was strongly warm from about 1925 until 1945, cool with brief warm episodes from 1945 until 1975, strongly warm from 1975 until 1996, and cool since then. These periods coincide rather neatly with warm and cool climatic periods and the PDO cycle offers an alternative explanation of climate change to variations in greenhouse gases.

A change in the sea surface temperature in the North Pacific might affect the global climate by slightly altering the extent of cloud cover. Clouds reflect sunlight, a property that makes them the principal atmospheric control of surface temperatures. As we all know, cloudy days are cooler than sunny days.

There is an alternative, or perhaps complementary, explanation for varying cloudiness, and it centres on sunspots. Sunspots are dark areas on the solar surface where the temperature is cooler than that of the surrounding area. Magnetic fields beneath the surface produce the spots and their number increases and decreases over a cycle of about 11 years. Solar astronomers have been studying them for thousands of years and there are long-term records of their cycles. In 1889 the German astronomer Gustav Spörer noted that very few sunspots were recorded between 1400 and 1510, a period when the climate was cold. This is now called the Spörer Minimum and its identification caught the attention of the English solar astronomer Edward Maunder. He checked the records and found another sunspot minimum from 1645 to 1715, which included a period of 32 years when not a single sunspot was reported. This is known as the Maunder Minimum and it also coincides with a period of cold weather, in fact one of the coldest parts of the Little Ice Age. Since then other minima have been discovered. John Dalton discovered the 1790–1820 Dalton Minimum, Johann Wolf the 1280–1340 Wolf Minimum, and Jan Hendrik Oort the 1010–1050 Oort Minimum. Each of these coincided very precisely with a climatic cold period. There have also been sunspot maxima, which coincided with warm periods.

How can the changing number of sunspots possibly affect the Earth's climates? No one really knows, but Henrik Svensmark, director of the Centre for Sun-Climate Research at DTU Space (the Danish National Space Center), and DTU Director Eigil Friis-Christensen have proposed a possible explanation. Earth is exposed to cosmic radiation, which consists of charged particles. Collisions between cosmic ray particles and air molecules cause sulphuric acid particles to form, on to which water vapour can condense to form cloud droplets. Cosmic radiation, therefore, contributes to cloud formation. The Sun emits a stream of charged particles, the solar wind. The solar wind deflects cosmic rays so fewer of them penetrate the atmosphere. During sunspot maxima, when the solar wind is strongest, fewer clouds form, and during sunspot minima, when the solar wind is weakest, cosmic rays are not deflected and cloudiness increases. If this is correct, it explains the link between sunspot minima and climatic cold periods and sunspot maxima and warm periods. But is it correct? Svensmark and Christensen have observational and experimental evidence to support their argument, but the jury is still out.

What is interesting is that a recent solar maximum, which peaked in about 1950, was the strongest for 400 years and that we are now entering a period of low solar activity, with few sunspots. Some astronomers suspect we may be heading toward a new sunspot minimum that will lead to a period of cold conditions between about 2030 and 2050.

Will the future be warmer or cooler, wetter or drier?
In December 2013 and January 2014 a total of 274 mm (10.8 in.) of rain fell over England, causing serious flooding in the Somerset Levels and the Thames Valley. As with every episode of extreme weather, environmental campaigners, journalists, and even some scientists were quick to claim the high rainfall as evidence of climate change. But was it? After all, this was not the first wet winter. In 1914 heavy rain began falling in November and continued through January, and in December and January delivered 276 mm. The same two months in 1929–30 were also wet, with 280 mm, and that wet period continued for three months and was much wetter than 2013–14. Since 1910 there have been eight two-month periods that were wetter than those in 2013–14.

In addition to the rain, the 2013–14 winter brought violent windstorms, which caused coastal damage. Were these something new? A 2012 report from the UK Met Office stated that "severe windstorms around the UK have become more frequent in the past few decades, although not above that seen in the 1920s. . . . There continues to be little evidence that the recent increase in

Deep waves in the polar jet stream brought heavy rain and severe gales that battered the southern coast of England, hurling seawater across seafront esplanades.

storminess over the UK is related to man-made climate change." The sea waves washed away the ballast from beneath a section of the London to Penzance rail track at Dawlish, in South Devon, leaving the rails suspended in thin air, swaying gently in the wind. This was not the first time that section of track had been washed away, however. It also happened in 1853.

Heavy though the rain undoubtedly was, neither the rain nor the ferocious winds can be safely linked to climate change. Indeed, in November 2013 the Met Office three-month outlook rated the likelihood of a drier than usual December–February at 25% and a wetter winter at 15%.

The winter weather of 2013–14 was exceptional but it was not unprecedented, and it was weather, not climate. What caused it? It was an unusually strong jet stream with very large Rossby waves in it that dragged a long series of midlatitude depressions across the North Atlantic, but although there was much speculation, no one understood why the jet stream behaved in that way.

So will the future be wetter than the past or drier, warmer or cooler? In its Fifth Assessment Report (AR5), published in 2013, the Intergovernmental Panel on Climate Change (IPCC) estimated that during the twenty-first century the mean global temperature will rise by 1.5–4.5°C (2.7–8.1°F) compared to the 1850–1900 average, and sea levels will rise by 0.4–0.7 metres. The report acknowledged the pause in the rise in temperature and based more of its conclusions on expert assessments than on the climate models it had used in previous reports.

The IPCC considers changes in the atmospheric concentration of carbon dioxide as the principal factor controlling temperature. Temperature changes therefore centre on estimates of the equilibrium climate sensitivity (ECS), which is the response of the global mean surface air temperature to a doubling of the carbon dioxide concentration. AR5 estimates this as 1.5–4.5°C, but the majority of observations of temperature and carbon dioxide concentration lead to calculations of an ECS value lower than 1.5°C. This means that past predictions of future warming have been consistently too warm. The atmospheric temperature appears to be less sensitive to changes in carbon dioxide than had been supposed, and in this section of its report the IPCC is markedly less confident than it had been in its earlier reports. It is also less confident that there will be more extreme weather. It thought it likely that since about 1950 heat waves had become more frequent in Eurasia and Australia, and that droughts had become less frequent in North America, but it scaled back its earlier warnings of extreme storms, floods, and droughts.

The hiatus in warming and doubts over the ECS indicate that while greenhouse gases certainly affect the temperature, other factors are also involved, and it is possible that those factors might overwhelm the greenhouse effect. Some of these factors involve natural climate cycles.

The Atlantic Multidecadal Oscillation (AMO) produces alternately warm and cool temperatures in the North Atlantic on a timescale of decades, very like the Pacific Decadal Oscillation (PDO). During the 1930s and 1960s to 1970s both cycles were synchronized, their warm and cool phases occurring together. Around 2000 they were out of phase. The PDO had entered its cool phase, which may have been sufficient to check the rising temperature, but the AMO was still in its warm phase. It is probable that the AMO will remain in its warm phase until some time in the 2020s. If it then switches to its cool phase and the PDO also remains in its cool phase, it is possible that global temperatures will fall.

There is also the possibility that we are entering a sunspot minimum, perhaps comparable to the Maunder Minimum. If so, it is distinctly possible that global temperatures will fall, though perhaps not so far as they did during the Little Ice Age because atmospheric carbon dioxide levels will continue to rise, exerting an offsetting warming influence.

We find ourselves on familiar ground. In common with our ancestors throughout the whole of history, we cannot tell what the world's climates will be like decades from now and far less centuries from now. No one can predict the future, but perhaps there are a few clues.

Maybe the warming of the late twentieth century will resume some time within the next decade or two and then continue. In that case we may expect a modest rise in temperature by the end of the century. This will lead to increased crop yields through the combined effect of warmer temperatures and higher levels of carbon dioxide, but not to any marked change in the severity or frequency of extreme weather events.

Alternatively, if it is the PDO and AMO that dominate the climate, temperatures may remain fairly constant until the 2020s and then begin falling. The resulting cool period will last for several decades before temperatures start rising once more.

Finally, if the relationship between cosmic rays and cloudiness controls the weather, and we are entering a sunspot minimum, we can anticipate a more prolonged cool period.

Attempts to control the weather

Looking back over the years, it seems that for most of our lives the weather has remained more or less constant. We remember especially cold winters and the long, happy days of those distant glorious summers of childhood, but the good and bad pretty much balance out, and we know our trickster memories preserve the exceptional and lose the ordinary. By and large, the climate stays the same.

It is hardly surprising, therefore, that the realization that the climate as a whole is changing seems alarming, and the alarm comes tinged with guilt when we are told that we ourselves are responsible for the change. Believing all change to be for the worse, we hasten to seek remedies. Having urged one another with only modest success to alter radically and expensively the very basis of the world's economies, some have turned to geoengineering, which is intervening physically to modify the way the climate works. Giant sunshades have been suggested, launched into orbit about the Earth to reflect some of the sunlight back into space. There are those who would capture atmospheric carbon dioxide and imprison it where it can do no harm. Iron has been added experimentally to the oceans to stimulate the proliferation of phytoplankton that will draw carbon dioxide out of the air by photosynthesis, and deposit it in the deep ocean as these tiny organisms die and sink. In an earlier episode

This light aircraft will burn a compound that releases smoke, which will condense to form silver iodide crystals in the hope of seeding clouds and producing rain.

of panic, when we were warned that the climate was growing colder, someone suggested coating the world's ice caps with soot, so the black material would absorb sunlight and raise the temperature. It seemed a good idea until the penny dropped that the first fall of snow would blanket and hide the soot.

If humans are already altering the climate then perhaps it is reasonable to assume we are capable of taking proper control and that we might produce a more congenial climate, or at least improve some aspects of day to day weather. The idea is far from new.

On 1–3 July 1863, the Battle of Gettysburg was fought beneath clear blue skies. Then, on 4 July, it began to rain. It rained and rained until the roads turned to quagmires. Had the battle caused the deluge? Again, the idea had been voiced before. In his *Life of Marius* (vol. IX of *The Parallel Lives*), Plutarch wrote: "And it is said that extraordinary rains generally dash down after great battles, whether it is that some divine power drenches and hallows the ground with purifying waters from Heaven, or that the blood and putrefying matter send up a moist and heavy vapour which condenses the air, this being easily moved and readily changed to the highest degree by the slightest cause."

Is it true? It very likely is, because commanders prefer to make battle on firm, dry ground and spells of fine weather end with rain. At all events, in 1891 the United States Congress appropriated $9000 to pay for experiments in which explosives carried aloft by kites and balloons were detonated inside clouds and cannons were fired into clouds to see whether this would induce the clouds to give up their rain. The ordnance failed to deliver. You can't explode rain out of the sky.

Maybe there's another way to make it rain. Vincent Schaefer was a toolmaker who constructed machines and laboratory devices as a member of a research team led by Irving Langmuir at the General Electric Research Laboratory. The twelfth of July, 1946, was very hot and Schaefer was trying to maintain a temperature of –23°C (–9°F) inside a box he was using as a cloud chamber. He dropped some dry ice—solid carbon dioxide—into the box and instantly millions of ice crystals formed from the supercooled water droplets in the cloud chamber. Ice crystals are a first step in the formation of raindrops and later Schaefer's colleague Bernard Vonnegut found he could produce them by burning silver iodide and feeding the smoke into the cloud chamber. Amid great publicity and assiduous promotion from Langmuir, on 13 November Schaefer dropped dry ice from an aircraft into a cloud over Pittsfield, Massachusetts, and snow fell in a line along the aircraft's track. By the 1950s commercial companies were offering rainmaking services, but using silver iodide. Other materials were also tried. Did any of it work? It is difficult to say. When rain did fall, who knows whether it would have fallen anyway? The craze passed and no one believes in it anymore.

Before it died, rainmaking led to more ambitious schemes, including several that attempted to modify hurricanes. In 1947, under Project Cirrus, Schaefer dropped dry ice into Hurricane King, about 640 kilometres from Orlando, Florida, and expected to head out to sea. Whether this had any effect is unclear, because unfortunately the hurricane altered course, made landfall near Savannah, and caused one death and more than $23 million in damage. Not deterred, between 1962 and 1983 Project Stormfury made repeated attempts to influence hurricanes. It had no success with hurricanes, but Cuba accused the United States of driving a fierce hurricane on to their territory, and Mexico denounced the United States for causing a severe drought.

Fog was a problem for air forces during World War I, and after the war ingenious inventors devised a means of dispersing it by heating runways. Their research intensified in the early years of World War II, leading to the formation in Britain of the Petroleum Warfare Department and Fog Intensive Dispersal Operation (FIDO). FIDO lined runways on both sides with burners

A World War II Lancaster bomber landing in an airfield from which the fog has been dispersed by FIDO, the Fog Intensive Dispersal Operation.

Smudging releases smoke and gases that absorb radiation from the surface. This helps prevent the temperature from falling too far on clear nights when there is a risk of frost.

fed with petrol by pipes from a reservoir. When ignited, the flares heated the air sufficiently to vaporize the fog, allowing aircraft to land safely (taking off in poor visibility using instruments was never a problem). FIDO worked and allowed Allied aircraft to operate from UK bases when German aircraft were grounded. But it was prohibitively expensive and eventually rendered obsolete by the introduction of radar guidance for landing.

The United States military conducted a number of experiments in weather modification in Southeast Asia between 1966 and 1972. Based on the notion that it is possible to control the weather, these attempts joined those in bacteriological, chemical, and other forms of environmental warfare that strongly influenced the belief among environmentalists that the environment is inherently fragile and threatened by our activities.

There are at least two versions of weather control that survive to the present. Many fruit growers use smudge pots to protect blossoms from late frosts. The smudge pot burns oil to produce smoke, carbon dioxide, water vapour, and other substances that together absorb infrared radiation, thereby reducing the loss of heat by radiation into a clear sky.

The other survivor is the evidently undying hope that the weather can be blasted into submission. Until the middle of the eighteenth century people rang storm bells, prayed, and performed rituals to avert bad weather. The alternative was to attack the weather. In the Middle Ages archers were employed to fire into storm clouds to prevent hailstorms and more recently warships fired cannons to dissipate storm clouds at sea. Austrian farmers used to fire guns into

A hail cannon that will be fired into any threatening cloud in the hope of protecting the crop from hail damage.

storm clouds to prevent hailstorms, and in 1896 Albert Stiger, a vine grower who was suffering losses from hail damage, took to firing cannons of his own construction into the clouds. This idea caught on and an international conference on hail shooting was held in 1901. The cannons themselves were conical, like large megaphones, mounted on sturdy wooden bases, some with wheels. In some places lookouts were stationed on high ground to warn of approaching storms in time for the gunners to man their posts. Hail cannons died out in Europe, but they are still used on the Great Plains of the United States. Do they work? It seems unlikely, but true believers swear by them.

Weather

Climate is a given. If you live in Greenland you know that most of the time the weather will be cold. It would be possible to grow bananas, but you would need a heated greenhouse with shading to provide darkness on summer nights and lighting to simulate daylight in winter.

You could do it, but it would be much cheaper to buy imported bananas from the supermarket.

Weather is less predictable. It comprises the events we encounter every day and it is time to think about what causes them. Why does the wind blow? How do clouds form and why do only some clouds produce rain? What causes hail, blizzards, thunderstorms, and even more violent assaults on our homes and gardens?

Why the wind blows

In the course of the evening TV weather forecast, the forecaster usually refers to the isobar chart. This is a map like the one in the illustration opposite, showing the distribution of surface atmospheric pressure across a geographic area. Sometimes, as shown here, the isobars are labelled with the pressure, usually in millibars, but labelling is not important. The point the forecaster is making is that the closer together the isobars are, the stronger the winds will be, and the winds will blow almost, but not quite, parallel to the isobars. So what's going on?

Air is a fluid, free to move where it will, but always under the control of gravity. Being a physical substance, air has mass, which means that in the Earth's gravitational field it is possible to weigh a volume of air and its weight will vary according to its density. The denser it is, the more molecules that will be packed into the volume, so the more it will weigh. At average surface pressure and temperature, 1 cubic metre of air weighs 1.23 kilograms. Its density decreases rapidly with increasing height, because air is easily compressed. Air near the surface is compressed by the weight of the column of air above it, but the overlying column becomes shorter with increasing height, and the pressure exerted by its weight decreases.

If you press down on the top of a column of a fluid you would expect the fluid to be squeezed out of the column at the bottom. Air cannot squeeze out, however, if the column is contained in a much larger column of air with the same density. But it can squeeze out if there is adjacent air with a lower density and it will try to do so until the two bodies of air have the same density.

Instead of density, let's describe the bodies of air in terms of the pressure exerted on them by the overlying air. Denser air forms a region of comparatively high pressure and less dense air a region of low pressure. A traveller moving from the centre of one to the centre of the other will observe a steady change in pressure, provided, that is, she remembered to bring a suitable barometer. The rate of pressure change will depend on the difference in pressure between the two centres and the distance between them, and if our

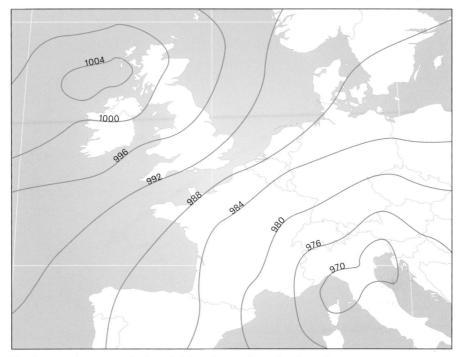

The lines on the map are isobars, joining places of equal surface air pressure. The numbers show the pressure in millibars on each isobar.

traveller also remembered the map, she could plot the changing pressure by drawing isobars.

Now let's do something strange. Let's discard the map and the traveller and imagine an isobaric surface where the pressure is the same everywhere. Regions where the pressure is higher protrude upward on such a surface, and regions of low pressure form depressions. The variations in pressure now resemble hills and valleys, the isobars resemble contour lines, and the distance between isobars marks the steepness of a pressure gradient, like the side of a hill. The diagram on page 108 shows a series of labelled isobars and the pressure gradient they denote.

Balance a football on the top of a hill and give it a nudge and the ball will roll down the hill. Air is not a solid, however. You can't balance anything on top of the "hills" on an isobaric surface. Rather, the entire hill is trying to move into the nearest valley and fill it. Left to its own devices, the air would find its own level and there would be no hills or valleys, just a level pressure plain (and that would be the end of all our weather). The difference in pressure exerts a

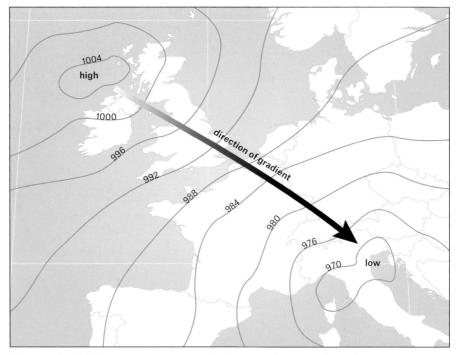

The rate at which the air pressure changes between regions of high and low pressure varies according to the distance between the two centres and the difference in pressure between them. Isobars therefore resemble contour lines on a topographic map.

pressure pushing the air at higher pressure to flow toward the region of lower pressure. The magnitude of the force is proportional to the pressure gradient and it is called the pressure-gradient force (PGF). It is given by:

$$F_{pg} = (1/d)(\Delta p/\Delta n)$$

where F_{pg} is the pressure-gradient force, d is the density of the air in kilograms per cubic metre, Δp is the difference in pressure in pascals, and Δn is the distance between the two centres in metres.

Pushed by the PGF, obviously the denser air will flow down the pressure gradient. So why doesn't it? It doesn't because the air lies above a rotating planet. As the air moves, the surface beneath it is also moving eastward by 15° of longitude an hour, which works out at 1668 km/h at the equator, with the result that moving objects appear to be deflected to the right in the Northern Hemisphere and to the left in the Southern Hemisphere. This phenomenon

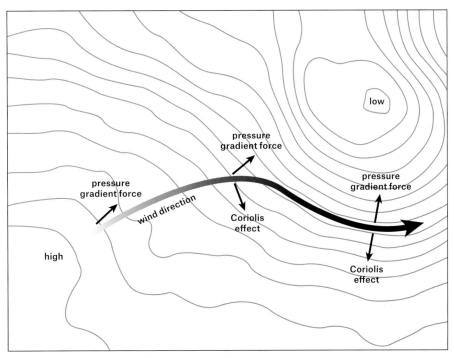

The pressure-gradient force (PGF) acts at right angles to the isobars. The Coriolis effect (CorF) deflects the moving air to the right (in the Northern Hemisphere). When the PGF and CorF balance the air flows parallel to the isobars.

had been known for centuries, but it was first explained in 1835 by the French physicist Gaspard-Gustave de Coriolis and came to be called the Coriolis force (CorF). When scientists realized that it is not really a force, because nothing is actually pushing the moving body to the side, the name was changed to the Coriolis effect, but the abbreviation CorF remained.

Meridians—lines of longitude—converge at the poles. Consequently, the change in their orientation over a given distance is greater at a high latitude than at a low latitude. This means that the magnitude of the Coriolis effect increases with latitude. It also varies according to the speed of the moving body. The magnitude is given by $2 \, \Omega \sin \phi v$, where Ω is the angular velocity of the Earth (7.29×10^{-5} radians per second), ϕ is the latitude, and v is the speed. The equator is at latitude $0°$, so when $\phi = 0$, $\sin \phi = 0$, and the poles are at latitude $90°$, and $\sin 90 = 1$, so when $\phi = 90$, $\sin \phi = 1$. Therefore, the Coriolis effect is zero at the equator and at a maximum at the poles.

As air moves under a PGF acting at right angles to the isobars, the CorF deflects it to the right (in the Northern Hemisphere) and, as the diagram on page 109 shows, when the PGF and CorF are in balance, the resulting flow of air—the wind—flows parallel to the isobars.

So that is why the wind blows parallel to the isobars. At least, it blows nearly parallel to the isobars. The geostrophic wind that really blows parallel to the isobars occurs well above the surface. Low-level wind is affected by friction due to the uneven surface. Friction slows it, reducing the magnitude of the CorF and allowing the wind to blow across the isobars at an angle between about 10 and 30 degrees depending on the wind speed and the surface topography.

How temperature changes with altitude

After completing some necessary but exhausting task on a warm summer afternoon, have you ever laid back to rest on a recliner and found yourself gazing lazily at the fleecy little clouds that drift sleepily across the sky? They are called fair-weather cumulus, but probably you knew that. Have you ever found yourself wondering why they are at a particular height, and all at the same height? Seen from the recliner their bases and tops are very clearly defined giving the clouds a look of solidity. They are not solid, of course, and if you were to study them through powerful binoculars you would see that their edges are much less sharp than they seem. Wisps of cloud are constantly breaking away from them and dissipating.

Clouds are made from tiny droplets of water, ice crystals, or very often a mixture of the two, and they form when water vapour condenses. The amount of water vapour present in the air varies greatly from place to place and from time to time and it is measured as the humidity of the air. There are several ways to report humidity, but the most widely used one is relative humidity (RH), which is the amount of water vapour in air at a specified temperature expressed as a percentage of the amount needed to saturate the air at that temperature. Cloud droplets or ice crystals will form, therefore, when the RH reaches 100%, provided there are small solid or liquid particles on to which the vapour can condense. The amount of water vapour air can hold is proportional to the temperature. Warm air is able to hold more water vapour than cool air can. Consequently, if the air temperature falls the RH rises, and if the temperature rises the RH falls, all without adding or removing any water vapour at all.

In order to persuade water vapour to condense, therefore, all we need do is lower the temperature—and if air rises, this will happen perfectly naturally. Everyone knows you need to take a warm sweater if you're setting off for a hike in the hills because it will be colder up there than it is down in the valley.

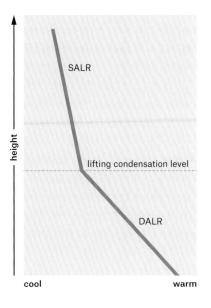

height

cool   warm

SALR

lifting condensation level

DALR

The lapse rate is the rate at which air temperature changes with height. There are several lapse rates, the most widely used being the dry adiabatic lapse rate (DALR) and saturated adiabatic lapse rate (SALR). The DALR changes to the SALR at the lifting condensation level.

Air temperature decreases with height. Set off from sea level on a day when the temperature is 15°C (59°F) and by the time you've climbed to about 2500 metres it's likely to be just below freezing.

The rate at which the temperature decreases, or lapses, with altitude is known as the lapse rate, and there isn't just one lapse rate, there are several. The first is the actual rate, measured in a particular place at a particular time. It is known as the environmental lapse rate (ELR) and it is constantly changing. If the temperature at sea level is, say, 15°C, the temperature at the tropopause is –55°C (–67°F), and the tropopause is at a height of 11 kilometres, the ELR is about 6.36°C (11.45°F) per kilometre.

Imagine now what happens to air that rises. As it rises the density of the surrounding air decreases, because air density decreases with height. The decreasing density allows the rising air to expand, its molecules moving farther apart, but in doing so the molecules expend some energy, which makes them slow down, a change that a thermometer registers as a fall in temperature. This fall in temperature occurs without any exchange of energy with the surrounding air, and it is said to be adiabatic. If the rising air is unsaturated, the rate at which its temperature decreases as it rises—and increases as it subsides and is compressed, gaining energy—is called the dry adiabatic lapse rate (DALR) and it is 9.8°C (17.6°F) per kilometre.

If the rising air is saturated, however, its water vapour will be condensing and condensation releases latent heat, which warms the air, so its temperature will fall more slowly than that of dry air. It will cool at the saturated adiabatic lapse rate (SALR) and if the saturated air subsides it will also warm more slowly than dry air because as compression raises its temperature its water will vaporize, absorbing latent heat. The SALR varies with the air temperature from about 3°C per kilometre in very warm air to about 9°C/km in very cold air, with an average of about 6°C/km.

Rising dry air cools at the DALR and as its temperature falls its relative humidity rises. When the RH reaches 100%, water vapour will start to condense and the air will cool at the SALR. The height at which the change occurs, and condensation commences, is known as the lifting condensation level. The diagram on page 111 shows the effect, as the steep DALR changes to the shallower SALR.

What happens next? That depends on the relative values of the ELR, DALR, and SALR. If the ELR is greater than the DALR, then rising air will cool more slowly than the surrounding air, regardless of whether it is dry or saturated—because the DALR is greater than the SALR. At every height, therefore, the rising air will be warmer than the surrounding air, so it will continue rising. The rising air is then said to be absolutely unstable.

If the ELR is smaller than the SALR, on the other hand, rising air, be it dry or saturated, will cool faster than the surrounding air. It will soon reach a height where its density is equal to that of its surroundings so it will rise no higher and may subside. The rising air is then said to be absolutely stable.

The most common situation, however, lies between these extremes, and occurs when the ELR is smaller than the DALR but greater than the SALR, as the illustration opposite shows. Rising dry air will cool faster than the surrounding air, but if it reaches the lifting condensation level before its density matches that of its surroundings it will start to cool at the smaller SALR. This is smaller than the ELR, so the air will continue rising. The rising air is stable while it cools at the DALR, but becomes unstable once it starts cooling at the SALR. Its becoming unstable is conditional on it rising, for whatever reason, as high as the lifting condensation level. The air is then said to be conditionally unstable. Giant storm clouds form in conditionally unstable air.

Clouds and precipitation

When rising air reaches the lifting condensation level the water vapour it contains begins to condense. At least, it condenses provided there are suitable surfaces on to which it can condense. These are minute particles known as cloud

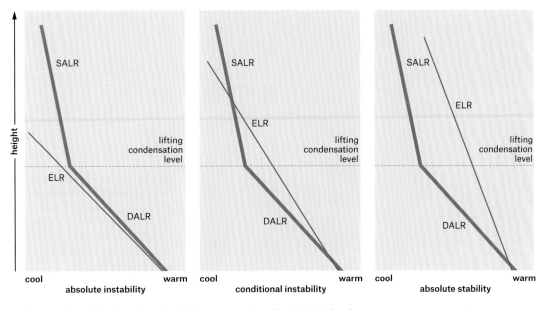

Absolute instability is when the **ELR** is greater than the **SALR**. Absolute stability is when the **ELR** is less than the **DALR**. Conditional instability is when the **ELR** is less than the **SALR** but greater than the **DALR**.

condensation nuclei (CCN). The most effective are 0.2–2.0 µm across and over land a cubic metre of air contains five to six billion. Air over the open ocean far from land contains about one billion per cubic metre. In very clean air, where CCN are in short supply, the relative humidity of the air can exceed 100% without the water vapour condensing. The air is then supersaturated.

Many CCN are hygroscopic—they absorb water and dissolve into it. Salt crystals make the best CCN. These enter the air when the water evaporates from drops of sea spray, and water vapour will condense on to them at a relative humidity below 80%. Many dust and smoke particles are also hygroscopic, as are particles of sulphate.

Condensation and evaporation go together and cloud formation is a dynamic process. Cloud droplets range in size from less than 1 µm to about 50 µm depending on the size of the CCN on which they form, but few last for more than an hour before they evaporate. The cloud is able to grow because droplets are forming faster than they are evaporating. A typical cloud contains about 100 million droplets per cubic metre and the largest droplets fall at about 25 centimetres per second.

If the air temperature is below freezing, cloud droplets may form ice

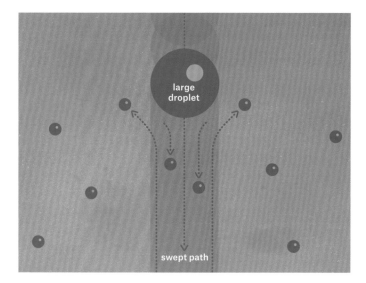

large
droplet

swept path

A large droplet falling through a cloud of small droplets collides and coalesces with those close to the centre of its path. Those farther from the centre are swept aside by the air displaced by the falling droplet.

crystals. For that to happen, freezing nuclei must be present. These must be similar in shape to ice crystals, and many freezing nuclei are splinters of ice that have broken away from aggregations of crystals. Others are fine clay particles. Freezing nuclei are less numerous than CCN. Air seldom contains more than about 100 per cubic metre. In the absence of freezing nuclei, the temperature of water vapour can fall well below freezing before it forms ice crystals spontaneously—under laboratory condition to −40°C (−40°F)—and even where they are present ice does not start forming until the temperature falls to about −10°C (14°F). At these temperatures the droplets are supercooled and if they strike a surface they will freeze instantly.

Warm clouds consist of liquid droplets, cold clouds of ice crystals, and mixed clouds contain both at different levels. Those wisps of cloud that close inspection reveals breaking away from the main body of a cloud enter unsaturated air and evaporate, and air that rises above the cloud also enters drier air, so its moisture no longer condenses, but air around the edges is constantly mixing between the cloud and its surroundings by a process called entrainment. A cloud, then, is a body of saturated air within a much larger mass of unsaturated air.

Cloud droplets and ice crystals are tiny, but in some clouds they succeed in merging to form raindrops or snowflakes that are heavy enough to fall from the cloud. There are two ways this can happen. In a mixed cloud, where ice crystals and supercooled droplets are jumbled together, the droplets evaporate and the water vapour is deposited directly as ice on to the ice crystals, so the ice crystals grow at the expense of the supercooled droplets. As the crystals grow, they collide with others and merge, eventually forming aggregations large enough to be called snowflakes that fall through the cloud. If the lower region of the cloud and the air beneath are above freezing temperature, the snowflakes will melt and fall as rain. In middle latitudes, most of the rain that falls, even in summer, is melted snow. If the lower part of the cloud is below freezing the snowflakes will survive and they will reach the ground as snow if the air temperature beneath the cloud is below about 4°C (39°F). Whether or not the snow settles depends on the temperature of the ground surface.

Droplets in warm clouds grow by collision and coalescence. All droplets will fall in still air, but the speed at which they fall depends on their size or, to be more precise, on the ratio of their volume to their surface area. A small droplet has a larger surface area in relation to its volume than a large droplet has, so it will offer more resistance to the air. Consequently, large droplets fall faster. They fall through a cloud of small, widely dispersed droplets that are falling more slowly and they will coalesce with those they contact. Collisions will occur only with droplets close to the centre of the path of the large droplet. The air displaced by the falling droplet will push aside those farther from the centre. Behind the falling droplet, however, small droplets are drawn into the wake of the large droplet and merge with it. You can see a very similar process when raindrops trickle down a windowpane. The diagram opposite shows how a falling droplet sweeps a path through the smaller droplets.

Any liquid water or ice that originates in the air is known as precipitation. The term includes mist, fog, drizzle, rain, frost, hail, sleet, snow, and ice pellets. Drizzle comprises droplets smaller than 0.5 millimetres in size and rain of drops 2–5 millimetres. In Britain, sleet is a mixture of rain and snow falling together, in North America it consists of ice pellets formed from raindrops that freeze below the cloud base. Ice pellets are snowflakes that have partly melted and refrozen. They are smaller than about 5 millimetres.

For thousands of years scholars searched for a way to describe unambiguously the many types of clouds that appear and disappear above our heads. It's a lot more difficult than it may seem and it was not until 1803 that Luke Howard, a London industrial chemist and amateur meteorologist, devised a scheme that worked and that formed the basis of the classification system used today.

The modern system divides clouds by their appearance into ten genera. The genera are then arranged by the usual height of their bases as high, middle, or low clouds. The table on page 117 shows this arrangement.

Cirrus cloud.

Altocumulus cloud.

Grey, featureless, stratus cloud.

Fair-weather cumulus cloud.

Cumulonimbus cloud, the top swept forward by the wind to form an incus, or anvil.

Wave cloud (lenticularis) forms in air that rises to cross a mountain peak. It often forms curious shapes, as it has done here.

HEIGHT OF CLOUD BASE (KILOMETRES)

	Polar regions	Temperate latitudes	Tropics
High: cirrus, cirrostratus, cirrocumulus	3–8	5–13	5–18
Middle: altocumulus, altostratus, nimbostratus	2–4	2–7	2–8
Low: stratus, stratocumulus, cumulus, cumulonimbus	0–2	0–2	0–2

Cirrus (Ci) is composed wholly of ice crystals and appears as white patches or wispy filaments.

Cirrostratus (Cs), also composed of ice crystals, forms a thin, white veil through which the Sun and Moon are clearly visible.

Cirrocumulus (Cc), also composed of ice crystals, forms small patches or approximately spherical elements.

Altocumulus (Ac) is composed of liquid droplets and often appears as elements forming waves or lines, although it is very variable.

Altostratus (As), also formed from liquid droplets, forms a thin, uniform sheet or fibrous veil that may be thick enough to obscure the Sun and Moon.

Nimbostratus (Ns), formed from liquid droplets that may be mixed with ice crystals, is uniformly grey and often delivers steady, continuous rain or snow.

Stratus (St), composed of liquid droplets sometimes mixed with ice crystals, forms a uniform, grey layer that may deliver drizzle, fine snow, or ice crystals.

Stratocumulus (Sc) consists of liquid droplets and appears as rounded masses or rolls of dark cloud against paler cloud, often with gaps large enough for sunlight to shine through.

Cumulus (Cu), composed of liquid droplets, has a white, fleecy appearance and often forms separate clouds with blue sky between them, but it may also be immersed in larger cloud and hidden.

Cumulonimbus (Cb) is dense and often extends to a great height from a low base. It consists of both liquid droplets and ice crystals. The top is often flattened against the tropopause and swept out by the wind, forming an anvil shape as ice crystals fall from it and vaporize in the dry air. It is the cloud associated with heavy showers, hailstorms, thunderstorms, hurricanes, and tornadoes.

The ten cloud genera are further divided into 14 species: calvus, capillatus, castellanus, congestus, fibratus, floccus, fractus, humilis, lenticularis, mediocris, nebulosus, spisssatus, stratiformis, and uncinus.

There are also nine varieties: duplicatus, intortus, lacunosus, opacus, perlucidus, radiatus, translucidus, undulatus, and vertebratus.

In addition, accessory clouds, such as arcus, incus, mamma, and virga, may be linked to main or mother clouds.

Latent heat

Around 1760 the Scottish chemist Joseph Black had become interested in a new problem that he was uniquely equipped to investigate because of the extreme meticulousness with which he worked and the pains he took to ensure his measurements were accurate. What Black had noticed was that when he warmed ice while monitoring its temperature, he found that although the ice gradually melted, its temperature did not alter as long as some ice remained. Since he was applying heat without affecting temperature, he concluded that heat and temperature are not the same thing. There can be a quantity or amount of heat, and an intensity of heat. Temperature is one measure of heat's effect, not of heat itself. A thermometer measures heat intensity, but not the quantity of heat. So we report temperature in degrees but heat in quite different units. The scientific unit of heat (also of work and energy) is the joule.

Black showed that the melting ice was absorbing a quantity of heat, which he supposed must be mixed with the ice particles and hidden among them. He called this latent, or hidden heat, contrasting it with sensible heat—heat you can detect with your senses.

In 1764 his assistant, William Irvine, helped him measure the much larger amount of latent heat that is absorbed when liquid water evaporates. Black also found that an equal amount of latent heat is released when liquid water freezes and when water vapour condenses.

We often describe heat as a form of energy, but this is misleading. In the ordinary sense of the word, heat does not exist at all. There is radiant heat, such as the heat from the Sun or from an electric fire, but that is electromagnetic radiation at a wavelength beyond red in the spectrum. Substances can possess only two forms of energy, potential and kinetic. Kinetic energy is the energy of motion and potential energy is the energy the substance stores by virtue of its location. A book on a shelf possesses potential energy. If it falls from the shelf its potential energy becomes kinetic energy. Heat, then, is the kinetic energy of the atoms or molecules from which a substance is composed. If you expose a substance to the high kinetic energy of the molecules racing about in a flame or to radiant energy, that energy will be imparted to the molecules of the substance and their kinetic energy will increase as they move faster. Eventually they may move so fast as to break free from the bonds linking them to their neighbours, and when that happens a solid will melt, or a liquid will boil. The energy that must be absorbed to accelerate the molecules to the point where they break free is the latent heat.

All substances absorb and release latent heat, the amount varying according to the strength of the bonds holding their molecules together. The table lists a few common substances with their latent heats of fusion (freezing or melting) and vaporization (evaporation or condensation). The melting point and boiling point temperatures are closely linked to the latent heat, so the table includes those.

LATENT HEAT

Substance	Fusion (kJ/kg)	Melting point (°C)	Vaporization (kJ/kg)	Boiling point (°C)
ethyl alcohol	108	−114	855	78.3
ammonia	339	−75	1339	−33.34
turpentine			293	
water	334	0	2501	100

The latent heat of deposition and sublimation is equal to the sum of the latent heats of fusion and vaporization. In the case of water this is 2835 kilojoules per kilogram. Latent heat varies somewhat with temperature, so these values for water are those at 0°C (32°F).

It is easy now to see why the saturated adiabatic lapse rate is lower than the dry adiabatic lapse rate. With each kilogram of water that condenses, 2501 kJ of energy are released to warm the surrounding air, thereby reducing the rate at which the temperature of the rising air decreases. It also explains the difference between stable and unstable air. Unstable air contains a large amount of water vapour. As the water vapour condenses, the air warmed by the release of latent heat rises farther. This causes further condensation, further warming, and the air just goes on rising. And as the air rises, its moisture condensing all the time, the cloud grows ever larger until it hits the ceiling of the tropopause. It is then large enough to generate severe storms.

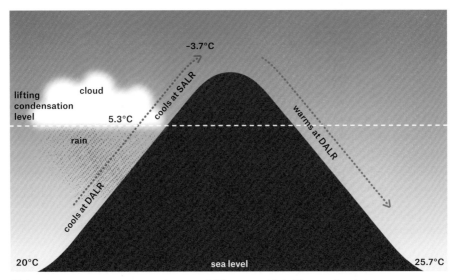

Moist air rises to cross a mountain range 3 kilometres high. The sea-level air temperature is 20°C. The rising air cools at the DALR until it reaches the lifting condensation level at 1.5 kilometres, where the temperature is 5.3°C. Cloud forms and rain falls. The air then cools at the SALR and crosses the summit at −3.7°C. The air is now unsaturated, so as it subsides it warms at the DALR, reaching sea level at 25.7°C.

Rain shadow

If you live on the lee side of a mountain range your climate will be much drier and somewhat warmer than that of the folks on the windward side. That is because you live in a rain shadow. Just how drier and warmer your weather is depends on the height of the mountain range and how far the mountains are from the coast. If you live in the centre of a continent the rain-shadow effect will be quite small, maybe too small to make much difference, but if the mountains are near the coast and the range is at right angles to the prevailing wind direction, then the effect will be dramatic. Several of the world's major deserts are in the rain shadows of mountains and form a distinctive type of desert.

Elsewhere, the difference is less spectacular. Scotland, for instance, receives its weather systems from the west, bringing moist air from the Atlantic, and the Highlands are a mountainous region occupying most of the northern part of the country. Oban, on the west coast, receives an annual average 1450 mm (57.1 in) of precipitation, while Edinburgh, in the east, receives 675 mm (26.6 in). The difference in climatic regimes continues through northern England, where the eastern and western sides of the country are separated by the Pennine hills. Manchester, in the west, receives significantly more

Steam fog is thin and wispy. It forms when cold air drifts across warm water and is often seen over lakes in winter.

precipitation (900 mm) than Leeds (693 mm), to the east of the Pennines.

Earlier, in discussing the way deserts form, we saw how air that is forced to cross high ground cools as it rises, loses moisture, and then warms and dries during its descent. The diagram opposite shows why this happens in more detail. Suppose there is a mountain 3 kilometres high and that it is situated in the path of the prevailing wind. Air approaches the mountain at sea level with a temperature of, say, 20°C (68°F). The air is forced to rise and as it rises its temperature decreases at the dry adiabatic lapse rate (DALR) of 9.8°C (17.6°F) per kilometre. Let's suppose further that the lifting condensation level is at 1.5 kilometres and the dewpoint temperature is 5.3°C (41.5°F). When the rising air reaches this height and temperature its water vapour will start to condense. Clouds will form and rain will fall on to the mountainside. The air, now saturated, will continue to rise, cooling at the saturated adiabatic lapse rate (SALR) of 6°C (10.8°F) per kilometre. When the air crosses the summit at 3 kilometres, therefore, its temperature will be –3.7°C (25.3°F).

By this point the air will have lost much of its moisture, so it is no longer saturated. Consequently, as it sinks down the lee side of the mountain it will warm by compression at the DALR and as its temperature rises its relative humidity will fall steadily. When it reaches sea level, therefore, the air will be dry and its temperature will be 25.7°C (78.2°F). Not only has the mere fact of crossing a mountain made the air drier, it has also made the air at sea level warmer on the lee side than on the windward side.

Fog

If you were to climb our imaginary mountain you might well walk into the cloud that blankets the slopes above the lifting condensation level. Surrounded by

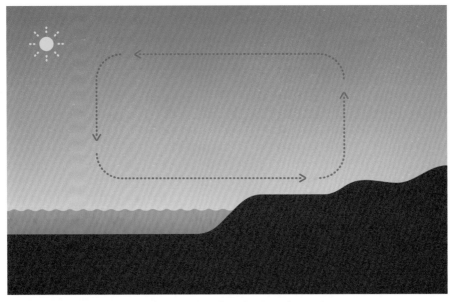

By day, warm air rises over land and cool air flows landward to replace it, as a sea breeze.

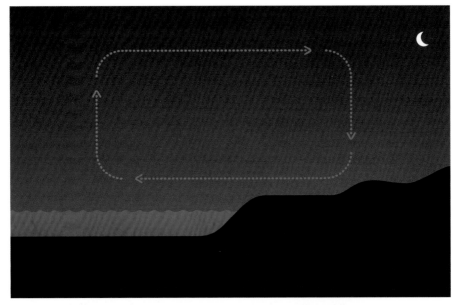

By night, air subsides over land as the surface temperature falls, and flows seaward as a land breeze, air from over the sea replacing it from above.

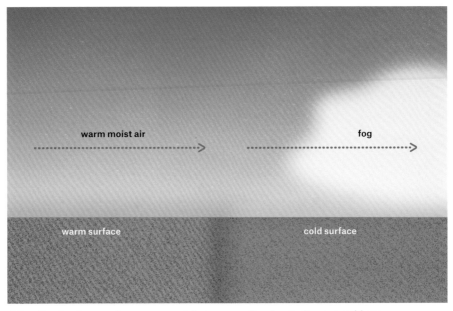

warm moist air

fog

warm surface

cold surface

Advection fog forms when warm, moist air moves (is advected) over a cold surface. Contact with the surface lowers the air temperature to below the dewpoint.

opaque whiteness that plays tricks with sound and drastically reduces the horizontal visibility, you might feel that you had encountered a bank of fog. And so you would. A person watching you from afar would see something apparently different, however. That observer would see you walking into a cloud. Both of you would be correct. Fog and cloud are the same thing, the distinction between them being based solely on whether or not they make contact with the surface. The climber is walking into hill or upslope fog. Seen from afar, it is stratus cloud.

By meteorological definition, a fog reduces horizontal visibility to less than 1 kilometre. If the visibility is greater than that, the cloud is called mist. Fog feels very wet, but typically 1 cubic metre contains less than 1 gram of water, in the form of droplets 1–20 μm in size, so small that they fall only very slowly. Nevertheless, where fog droplets coat the surfaces of projecting structures (such as branches) and drip to the ground—this is known as fog drip—they can deliver as much moisture as a light shower.

Fog can form either by cooling air to below its dewpoint temperature or by adding water vapour to air until the air is saturated. The steam in the bathroom is fog formed by releasing warm water vapour into cool air, causing the water vapour to condense. This also happens outdoors, when cold air moves across

All sea fogs are caused by advection. They are common in San Francisco Bay, in warm, moist air that crosses the cool water of the California Current.

the surface of water that is warmer. The fog is thin and wispy, and is often seen in winter over lakes. It looks like steam and is called steam fog.

In summer, fog occasionally forms in the afternoon over lakes and the sea. This is advection fog and it occurs when a wind of 10–30 km/h carries warm air across a cold surface. It most often develops in summer, when the land warms much faster than the water, so by afternoon the land surface is markedly warmer than the water surface, but it requires a breeze that blows from land to sea, which makes it fairly uncommon, because under these conditions it is more usual for air to rise over the warm land and for cool air flowing landward to take its place as a sea breeze by day, reversing to a land breeze by night, as the land cools more rapidly than the water. The illustration on page 123 (above) shows how sea and land breezes happen.

Onshore breezes can generate advection fog, however, when they carry warm, moist air across a cold ocean current. The illustration on page 123 (below) shows how this happens. All sea fogs form by advection and fogs of this type are common along the California coast in air approaching from the Pacific across the cold California Current. The most famous example is the fog that often rolls from the sea into San Francisco Bay, romantically shrouding the Golden Gates Bridge.

Less romantically perhaps, darkling beetles that live in the Namib Desert rely on the fog that often rolls in from the sea at night in air that has crossed

Radiation fog covers the ground in a shallow layer, seen here at dawn. As the morning advances and the temperature rises, the fog will evaporate from the ground upward.

the cold Benguela Current. Indeed, it is their only source of drinking water. On nights when the fog is coming, thousands of the beetles scramble to the tops of sand dunes and stand in rows on their long front legs so their hard, shiny elytra (wing cases) are upright and facing the coast. As the fog arrives, droplets coat the elytra and water trickles down to the insects' mouthparts.

The shallow layer of fog, seldom more than about 30 metres deep, that you often see early in the morning over rivers and wet fields forms in a very similar way. This is radiation fog and it requires moist air and fine weather, with a clear sky at night. With no cloud to absorb and reradiate long-wave radiation from the ground, the surface temperature falls sharply during the night. This chills the air in contact with the ground and fog forms when the temperature reaches the dewpoint. Radiation fog usually clears by midmorning. The sunshine raises the surface temperature and the fog evaporates, often from the bottom upward, because air warmed by contact with the ground rises by convection,

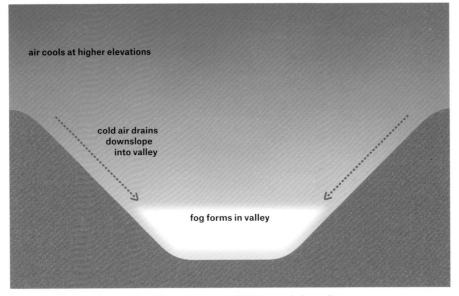

air cools at higher elevations

cold air drains
downslope
into valley

fog forms in valley

Valley fog forms at night when the surface on a hillside cools by radiation, chilling the air in contact with it. The cold air flows downhill into the valley where it cools further, to below the dewpoint temperature.

raising the temperature and therefore the relative humidity, in the fog. This can make it look as though the fog is lifting, although that isn't what is happening. You might also say that the fog is burning off, which is more accurate.

Valleys are especially prone to their own version of radiation fog. On clear nights the hillsides lose heat, chilling the air in contact with the surface. As the Sun sinks, the lower part of the valley sides are shaded first, and in steep-sided valleys the valley bottom may have been in shade all day. The boundary layer of cold air is denser than the warmer air above it and slides downhill, accumulating near the bottom. If the air is moist, fog forms as its temperature falls below the dewpoint. The diagram above shows how valley fog forms. Most radiation fogs clear quickly, but valley fog is the exception. As the early morning sunshine warms the sides of the valley, a layer of warm air lies above the fog, trapping it, and the fog reflects much of the sunshine falling on it, so its temperature remains low.

Fog droplets that coat surfaces can easily be wiped away. If the surfaces are below freezing temperature, however, the droplets will freeze instantly on contact with them. This is freezing fog, and the surfaces on to which it will freeze include car windshields, so freezing fog can be dangerous. The droplets themselves may be close to freezing temperature or supercooled.

Freezing fog is not to be confused with frozen fog, which is a much rarer

Hoar frost, forming beneath a clear sky, encases twigs, leaves, grass blades, and every other exposed surface in ice crystals, producing the enchanted landscape of a winter morning.

phenomenon. Frozen fog forms when supercooled water droplets freeze, forming ice crystals that continue to grow as water evaporates from the remaining supercooled droplets and is deposited on to the crystals. It is not long before the ice crystals are heavy enough to fall, so frozen fog is short-lived.

Frost

On a crisp winter morning when there is not a cloud in the sky, the world can seem enchanted, the plants, right down to the very blades of the grass, encased in glittering crystals of silver ice, the spider webs hanging in the hedge made entrancingly visible and thereby harmless to every passing insect. The ice makes everything it coats appear hoary (grey as with age), so it's called hoar frost and it's the most common type of frost.

Fearful for the wellbeing of the garden you might wonder whether the ice harms the plants, but you can be reassured, because it is quite harmless. The deposition of ice releases latent heat, warming the adjacent air, and the coating

Fern frost forms on a windowpane at night, when the glass is chilled well below freezing and drops of moisture on the inside cool to below freezing before starting to solidify in what then becomes a chain reaction producing fern-like patterns of crystals.

Rime ice coats plants when supercooled fog or drizzle freezes onto them.

of ice, which forms early in the evening, greatly restricts the radiation of heat from the underlying surfaces. The frost protects the plants it coats from further chilling.

Hoar frost forms on clear, cold nights when the air is moist. Exposed surfaces radiate warmth and their temperature falls rapidly, chilling the adjacent air to below the frost point, which is the temperature at which water vapour turns directly into ice. The relative humidity of the air then rises to more than 100% and water vapour deposits as ice crystals that grow on to each other, producing the irregular hoary appearance.

Another type of hoar frost will also form if the temperature falls below freezing more slowly. As the air temperature above the cooling surfaces falls below the dewpoint water condenses on to them as dew, but when the falling temperature passes the freezing point the dew freezes. This frost looks different and is sometimes called white dew. It is usually hard, transparent, and with rounded surfaces that retain the shapes of the original dewdrops.

Freezing dew can also form fern frost, once common but not seen now in homes that have central heating and double-glazing. During the evening, moisture in the warm air inside the room condenses on to the cold windowpane. At night, as the outside temperature falls, the temperature of the window also falls, eventually to below freezing, and so does the temperature of the liquid droplets on the inside. At first the supercooled droplets remain liquid, but after a time ice crystals start to form between them. Supercooled water

WEATHER

Sometimes called white dew, this type of hoar frost forms when dew freezes slowly. The frost is hard, transparent, smooth, and has rounded surfaces that reproduce the shapes of the original dewdrops.

then freezes on to the ice crystals, starting a chain reaction that spreads to cover the window with a pattern of ice resembling fern fronds.

In really cold weather rime ice may form. It is white and hard, with a rough surface, and it forms when supercooled droplets in fog or drizzle freeze on contact with cold surfaces. It can also form by the direct deposition of water vapour. Once freezing has started more ice crystals form on the frozen surface, creating the irregular structure. Rime ice sometimes grows into delicate, elaborate shapes, especially when there is a wind, when it develops only on the sides of objects that face into the wind. In these circumstances the ice can reach a considerable thickness.

We can enjoy all of these types of frost, for they are beautiful and harmless. But low temperatures can kill. If the air is very dry its temperature can fall far below freezing without frost forming, for want of moisture. The moisture inside plant tissues freezes, however, killing cells and leaving the surface of the plant blackened. This is a black or hard frost and plants need protection from it.

Air frost is also dangerous. This is the condition in which the air temperature is below freezing, and not just the air in the boundary layer next to plant surfaces. Ice crystals can form in the tissues of plants exposed to such low temperatures for any length of time.

Ground frost, on the other hand, causes few problems. When the weather presenter forecasts ground frost it means the ground temperature will fall below freezing, but the air temperature will remain above freezing. You may get up in the morning to a scene of hoar frost, but plants will survive. Ground frost forms on clear, windless nights, when surfaces lose heat by radiation but the air above the surface remains comparatively warm. It will not form on cloudy nights because the clouds reflect and absorb and reradiate the radiation from the surface, nor on windy nights because the slightest breeze will mix the air, preventing a layer of cold air from collecting at the surface.

It can sometimes happen on clear nights, especially in late spring, that built surfaces such as roads and drives, retain enough warmth for their surfaces to remain above freezing, but nearby cultivated ground freezes and hoar frost forms on plants. This is a grass frost.

During the frost season the weather forecast may include an estimate of the frost hazard. This is the likelihood that plants will suffer frost damage during the forecast period. It may take the form of an estimate of the probability of a killing frost—a drop in temperature capable of killing plants—at a specified place and date during the growing season, or the frequency of killing frosts at a particular place in previous years. Alternatively, it may consist of a set of dates for the earliest and latest frosts over a number of years.

Hail

Most hailstorms are unpleasant but fairly harmless. The hailstones are small, typically about the size of a dried pea or even smaller, and although a heavy storm may deposit a thick layer of hailstones, these soon melt. But there are exceptions and hailstones can be much bigger. Traditionally, big hailstones are described as being the size of everyday objects including pennies, walnuts, golf balls, tennis balls, and grapefruit. On 23 July 2010, a hailstone 20 centimetres in diameter, weighing 880 grams (2 lbs), fell in Vivian, South Dakota, and storms have been known to bury an area beneath up to a metre of hailstones. These are exceptional, of course, but a shower of hailstones the size of golf balls is not uncommon, and it will leave cars dented, skylights broken, greenhouse glass and plastic shattered, and plants flattened. No wonder that

Giant hailstones, about 6 centimetres across. Their size shows that they formed in a very large, violent storm cloud.

A hailstorm has left a hailswath covering much of this street with hailstones.

over the centuries rural communities have tried prayer, bell ringing, and firing explosives into clouds in order to protect their property and livelihoods, all, alas, to no avail. The best protection is probably a hail net, which resembles a fruit cage but is sufficiently robust to catch and hold falling hail.

Hail forms only inside large cumulonimbus storm clouds, which is why thunderstorms often accompany hailstorms. The bigger the storm cloud the larger the hailstones will be, but the freezing level needs to be lower than about 3 kilometres in a cloud up to 10 kilometres deep in order for the hailstones to have room to form and grow. That is why hailstorms are rare in the tropics. Nor are they common in very high latitudes, where the air temperature is below about –30°C (–22°F) in the lower regions of the cloud, because in these clouds there are too few of the supercooled water droplets from which hailstones are made. A cloud capable of producing a hailstorm often has a greenish tinge and with luck you may see the hail approaching as a column—called a hailshaft—below the cloud. Once fallen, an area partly covered by hailstones is a hailswath and a hailstreak is a strip of ground that is completely covered.

Hailstorms are most frequent and the hailstones largest in the interior of midlatitude continents. Central and southern Europe, China, and southern Australia suffer greatly and part of the Great Plains of the United States is known as Hail Alley because of the frequency and severity of its storms. Cheyenne, Wyoming, has nine or ten hailstorms a year.

The inside of a big cumulonimbus cloud is a violent place. Warm, moist, unstable air is drawn into the base and rises, the latent heat of condensation supplying the energy to keep it rising, often all the way to the tropopause, and generating gales that blow upward at up to 160 km/h or more. The diagram

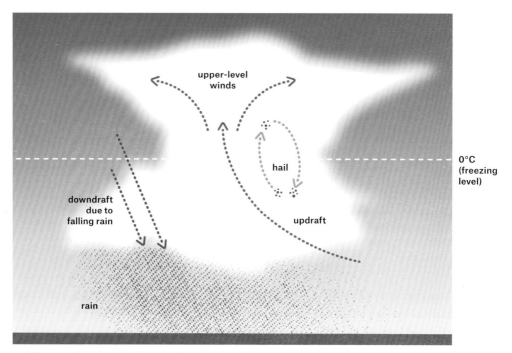

upper-level
winds

hail

0°C
(freezing
level)

downdraft
due to
falling rain

updraft

rain

Hail forms only in large cumulonimbus clouds, where strong updraughts
carry cloud droplets above the freezing level, where they form ice pellets.
The pellets fall, but re-enter updraughts, so they are carried up and down
repeatedly, all the time growing larger as more cloud droplets freeze on to
them, until they are too heavy for the updraughts to support them.

above illustrates the structure of such a storm cloud, greatly simplified because
it shows only one set of vertical currents. A supercell cloud has only one set of
currents, but they are aligned rather differently, and most storm clouds have
several sets. Multi-cell clouds often produce hailstorms, but it is supercell
clouds that grow the biggest hailstones.

Cloud droplets are swept upward in the updraughts and freeze when they
rise above the freezing level. They continue to rise as tiny ice pellets and in
their travels they encounter supercooled cloud droplets, some at just above
and some below freezing temperature. The warmer droplets coat the ice pellets with liquid water before they freeze, forming a layer of clear ice. The colder
droplets, occurring at a higher level, freeze instantly on contact with the hailstone, trapping tiny air bubbles in the crystalline structure and forming a layer
of white, opaque ice. Because of the way they form, hailstones often have an
onion-like structure of alternate clear and opaque layers. This structure is visible only on large hailstones, however. Small ones make too few circuits of the
cloud to acquire layers.

HAILSTONES AND UPDRAUGHT SPEED

Hailstone size (mm)	<6.4	6.4	13	18	19	22	25	32	38	44	51	64	70	76	101	114
Updraught speed (km/h)	<39	39	56	61	64	74	79	87	97	103	111	124	130	135	158	166

When it arrives in the upper region of the cloud where the updraughts weaken, the hailstone starts to fall, still growing as it descends. In the lower regions it enters another strong updraught and rises once more. This process continues, with the hailstone repeatedly rising and falling through the cloud of supercooled droplets, and growing larger all the time. Finally, it grows too large and heavy for the updraughts to support it, and its descent continues through the base of the cloud and to the ground.

Clearly, the stronger the vertical air currents the larger and heavier are the hailstones they can support, so it is possible to measure the size of the hailstones and use the measurement to estimate the speed of the air currents. The table above shows a range of hailstone sizes against the current speeds.

It is the large hailstones that cause all the damage and the Tornado and Storm Research Organisation (TORRO) has compiled a scale of hailstorm intensity relating the size of the hailstones to their kinetic energy and the severity of the damage they cause. The table on page 134 shows this categorization.

Meanwhile, toward the rear of the cloud, condensation produces raindrops. These are colder than the air below them and as they fall each drop drags with it a small envelope of cold air. In this way the falling rain produces a cold downdraught that flows out of the cloud and to the surface. That's why it feels cold in the rain.

Blizzards

Snow brings out the best and worst in us. Fallen overnight, it transforms the morning workaday world into an entirely new landscape, where no human foot has trodden but only passing deer have left their footprints. We play in it and rejoice at it and rummage for the family sledges and snowboards—until it's time to go to work. Then it's out with the snow shovel and ill-tempered

TORRO HAILSTORM INTENSITY SCALE

TORRO rating	Intensity category	Diameter (mm)	Likely kinetic energy (J/m^2)	Damage
H0	hard hail	5	0–20	none
H1	potentially damaging	5–15	>20	slight general damage to plants
H2	significant	10–20	>100	significant to fruit and plants
H3	severe	20–30	>300	severe to fruit and crops, damage to glass and plastic, paint and wood marked
H4	severe	25–40	>500	widespread to glass, vehicle bodywork
H5	destructive	30–50	>800	wholesale destruction of glass, damage to tiled roofs, risk of injury
H6	destructive	40–60		brick walls pitted
H7	destructive	50–75		severe to roofs, risk of serious injury
H8	destructive	60–90		severe, highest category ever recorded in Britain
H9	super hailstorm	75–100		extensive structural damage, risk of serious or fatal injury
H10	super hailstorm	>100		extensive structural damage, risk of severe or fatal injury

mutterings about heart attacks, pushing snow off windshields, blocked minor roads, late trains, and unreliable buses. How swiftly is delight transformed to misery.

The plants and small animals don't mind it. A layer of snow is like a blanket that protects them against the cold wind and provides a world where the mice and voles and shrews can scuttle about in search of food safely hidden from the hungry predators.

When the freezing level is below or at the same height as the lifting condensation level, the moisture in rising air will be deposited as ice onto freezing nuclei, and crystals will grow. Ice crystals, and the snowflakes they form, are hexagonal. But how does this six-sided shape emerge?

A water molecule comprises two atoms of hydrogen (H) bonded to one atom

The deep snow typical of early winter east of the Great Lakes is due to the lakes. Air that has been chilled as it crosses the cold lands of the Midwest warms and gathers moisture as it crosses the lakes, which are unfrozen. When it reaches the cold ground to the east, its moisture condenses and falls as snow. This is called the lake effect.

of oxygen (O) to make H_2O. As the illustration here shows, however, the molecule is asymmetrical. If you draw a line from the centre of each of the hydrogens to the centre of the oxygen the lines meet at an angle of 104.5 degrees. This arrangement leaves the molecule with a small positive charge on the side of the hydrogens and a small negative charge on the oxygen side. When water molecules join together they do so by bonds (called hydrogen bonds) between a hydrogen on one molecule and the oxygen of another. As the temperature falls the water molecules move closer together and hydrogen bonds form. Each molecule bonds to four others, but because of the shape of the molecule the hydrogen atoms can face toward two of those,

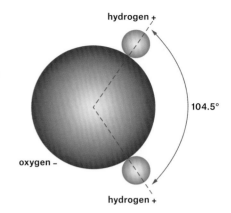

A water molecule comprises one atom of oxygen and two of hydrogen, the hydrogen atoms forming an angle of 104.5 degrees at the centre of the oxygen.

allowing each molecule to be oriented in one of six ways. Each of those six is equally likely, but the first to form forces molecules arriving later to orientate themselves in ways that maintain all of the hydrogen bonds. Consequently, although all ice crystals are basically hexagonal, they occur in a variety of hexagonal shapes.

There is an international system for classifying ice crystals. It recognizes ten basic shapes and allots a standard symbol, shown in the illustration opposite (above), for each. Plates are flat, hexagonal rings. Stellars are six-pointed rings. Columns are cylinders that are hexagonal in cross-section, and sometimes two or more columns are joined together. Capped columns are columns with a bar at each end. If they are joined together the bars remain intact. Needles are fine, like splinters, and may be joined together. Spatial dendrites have many fine branches, like fern fronds, and the fern frost that forms patterns on windowpanes is made from spatial dendrites. Irregular crystals are chaotic, with no regular shape. There are also symbols for graupel (soft hail), sleet, and hail.

Crystal formation is not random. The temperature and whether the air is saturated or supersaturated determine the shapes that occur. The graph opposite (below) plots the moisture content of the air, in grams per cubic metre, against temperature, with illustrations of the associated ice crystal types.

Once ice crystals have formed, they grow by deposition of moisture that evaporates from supercooled droplets, but the growth must maintain the overall hexagonal shape owing to the orientation of the hydrogen bonds. Snowflakes are the resulting aggregations of crystals. As a snowflake falls, it continues to

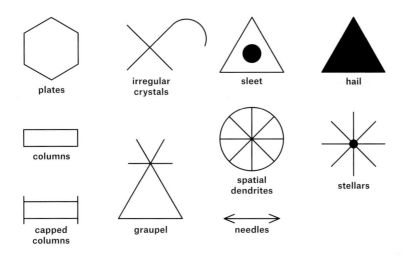

The standard symbols used to designate the ten recognized types of ice crystals.

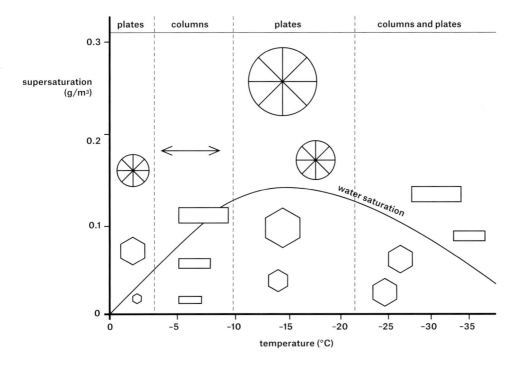

The conditions of supersaturation and temperature in which ice crystals form.

Blizzards often strike in cities, especially where tall buildings
funnel the wind along the streets.

grow, but it also loses fragments through collisions with other flakes, and it
may enter slightly warmer air and begin to melt, freezing again when it returns
to colder air. Each snowflake encounters slightly different conditions. That is
why, when they reach the ground where collectors can look at them, each snowflake appears to be unique. We should not make too much of this. Shapes do
repeat, but there is a truly vast range of possibilities.

Once they are below the cloud base, falling snowflakes are blown by the
wind. If the fall is heavy it can reduce visibility significantly, especially for
drivers because the flakes reflect car headlights. In extreme cases, falling snow
can effectively reduce visibility to zero, in a whiteout. In a whiteout, the sky, the
ground, the view in every direction, everything is white and, to make matters
worse, sounds are muffled and distorted. If you hear a sound, unless it's very
close you'll not be able to tell the direction or distance to its source. If you find
yourself in a whiteout while driving, don't abandon the car, but try to clear the
snow from it from time to time and prevent it from being buried. A car is much
easier for rescuers to find than a lone person. If on foot in the countryside, you

WINDCHILL

	0°C	-5°C	-10°C	-15°C	-20°C	-25°C	-30°C	-35°C	-40°C	-45°C
10 km/h	-3	-9	-15	-21	-27	-33	-39	-45	-51	-57
20 km/h	-5	-12	-18	-24	-31	-37	-43	-49	-56	-62
30 km/h	-7	-13	-20	-26	-33	-39	-46	-52	-59	-65
40 km/h	-7	-14	-21	-27	-34	-41	-48	-54	-61	-68
50 km/h	-8	-15	-22	-29	-35	-42	-49	-56	-63	-70
60 km/h	-9	-16	-23	-30	-37	-43	-50	-57	-64	-71

may need to dig a snow hole for shelter. On no account should you head off in search of help, because you will almost certainly quickly become lost.

Heavy falling snow can produce a whiteout, but it is not a blizzard. A blizzard is worse, and much more dangerous. It consists of wind-driven falling snow, snow blown up from the surface, or a mixture of both. Meteorologists define it as a wind of at least 55 km/h, a temperature no higher than –7°C (19°F), horizontal visibility of less than 400 metres, and if snow is falling it must be heavy enough to deposit a layer at least 250 millimetres deep. Persons caught in a blizzard rapidly become totally disoriented, surrounded by howling whiteness. And blizzards can happen anywhere, even in the most unlikely places. In February 1983 a blizzard killed 47 people in Lebanon.

In winter, weather presenters often mention the windchill factor. They mean that the wind will make it feel colder than you would expect at that air temperature, and they quote figures in degrees to indicate the extent of the chilling effect. This is misleading, because the air temperature does not change. Windchill exposes you to a temperature much closer to the actual temperature than you are used to. Our clothes, and a mammal's fur or a bird's feathers, trap a layer of air next to our skin and in the tiny air spaces in the fibres of our garments. Our body warmth maintains that layer at a comfortable temperature, but a strong wind can penetrate our clothes and sweep that layer away, replacing it with outside air. That is why the wind makes it feel colder than it is, the windchill temperature being equivalent to the temperature you would feel if fully clothed in still air. The windchill effect can be very dangerous. The table above shows the temperature you will feel at a range of air temperatures and wind speeds. Read the measured air temperature along the top row, the wind speed in the left column, and the apparent temperature where the two intersect.

The two main lightning channels seen here are about 20 centimetres wide. They are heating the air around them by up to 30,000°C.

Thunderstorms

It begins as a pure white cumulus cloud, but one that is growing rapidly into a full-blown cumulonimbus in very unstable air. Inside the cloud the vertical currents carry droplets above the freezing level where they become ice crystals, so the lower part of the cloud consists of liquid water and the upper part of supercooled water and ice. And in the violence of its growth, something strange is happening.

Droplets freeze from the outside, so briefly they form shells of ice enclosing liquid water and some of the water molecules are ionized. That is, they have lost or gained electrons and carry electric charge. The positive ions, which are the more mobile, migrate to the coldest region of the droplet, the negative ions remaining in the warmer centre, so as it freezes a droplet comes to consist of an ice shell with positive charge enclosing a liquid centre with negative charge. When water freezes it expands—think of the way it can burst water pipes—and as the core of the droplet freezes its expansion shatters the outer shell, releasing tiny ice fragments carrying positive charge. These are so small and light they accumulate in the upper part of the cloud, where the currents have carried them. The heavier core, carrying negative charge, sinks and the cores accumulate in the lower part of the cloud. Scientists are not entirely certain of all

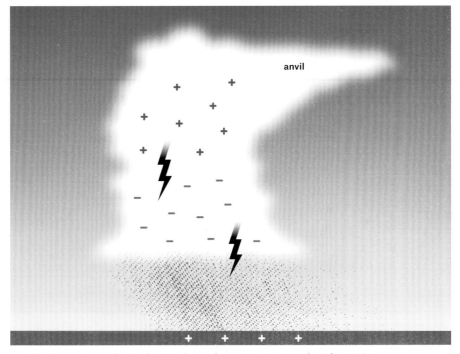

Inside a large storm cloud, electric charge becomes separated, with positive charge collecting at the top and negative charge lower down.

that is happening, but this much they understand and it results in a cloud that carries a positive charge in the upper part and a negative charge in the lower part, as the diagram above shows. The negative charge at the cloud base then induces a positive charge on the ground surface below the cloud.

An electric field then exists and as regions of dense charge develop the field starts to break down, the breakdown probably triggered by particles, released through the radioactive decay of naturally occurring atoms and cosmic radiation, which dislodge electrons from atmospheric gas molecules. The electric field accelerates the free electrons, which collide with more gas molecules, detaching more electrons that eventually become a runaway avalanche of electrons. These high-energy electrons then initiate lightning strokes that flash between regions of dense charge inside the cloud, between clouds, or between the cloud and the surface below. If the forked lightning stroke is hidden by cloud it will appear as a bright flash called sheet lightning.

A lightning stroke begins as an almost invisible stepped leader. This is a stream of electrons carrying negative charge that travels where the electric

This tree was split by lightning that heated its sap until the outer part of the tree exploded, blasting away the bark.

field offers least resistance, branching this way and that and ionizing the air it contacts to form a jagged lightning channel about 20 centimetres across. When it nears the positive charge, for example on the ground, the leader triggers a much more luminous return stroke carrying positive charge along the lightning channel.

The return stroke neutralizes the leader and the remaining charge in the cloud neutralizes the return stroke, and this triggers a second stroke, beginning higher in the cloud. This starts as a dart leader carrying negative charge and it is met by a return stroke, and this is repeated until the leaders and return strokes have neutralized the charge in part of the cloud. The entire process takes about 0.2 seconds, comprises three or four flashes about 50 milliseconds apart, and it is seen as a single flash.

Lightning strikes have different effects. If the leader reaches the ground or a tree and continues delivering charge until the return stroke, it will heat dry material enough to ignite it. This is called hot lightning and it is the type that starts forest, bush, heath, and grass fires. Cold lightning does not start fires because the charge delivery is not sustained until the return stroke, but it can cause damage by heating the sap beneath the bark of trees until it boils, causing the outer part of the tree to explode, blasting away the bark and sometimes bringing down the tree.

Most thunderstorms produce precipitation and usually it is heavy. There are exceptions, however, and dry thunderstorms produce what is known as dry lightning. Storms of this type occur most frequently in summer in the western United States. It is not only cumulonimbus clouds that produce lightning. It also occurs in clouds of ash and gas ejected in volcanic eruptions.

A lightning stroke carries a very large amount of energy that heats the air through which it travels. Inside the lightning channel the temperature rises by up to 30,000°C in less than a second. The sudden heating causes the air inside the channel to expand, increasing the pressure to 100 times its ordinary value, and the air, quite literally, explodes, releasing the shock waves that we hear as thunder.

Sound waves travel through air at about 1080 km/h and the light from the lightning flash travels about one million times faster, at the speed of light. Consequently, an observer some distance away will see the flash before hearing the sound. The difference in travel times means that a gap of three seconds between the flash and the sound represents a distance of 1 kilometre.

As the sound of the thunder travels, their passage through the air damps the sound waves. The high frequencies, carrying high pitch, disappear first, so the greater the distance to the flash the lower the pitch of the thunder will be, until it is nothing louder than a deep rumble that may continue for several seconds. It continues because a lightning stroke can be more than 1.5 kilometres long, so the sound waves have different distances to travel and we hear those from the nearest part of the flash first with the others following. Thunder is seldom audible more than about 10 kilometres away, because by then all of the sound waves have been either absorbed by the air or refracted upward.

Inside each of the cloud's convection cells there are air currents carrying warm air upward. It is these that drive the growth of the cloud and the development of the thunderstorm. But the precipitation drags cold air downward and eventually the cold downdraughts spread into the warm updraughts, chilling the rising air and in that way suppressing it. As less and less air is swept upward, there is less condensation and release of latent heat and the entire cloud loses its source of energy. It has reached the final stage in its life cycle and starts to dissipate. People sometimes say that the cloud has released all of its moisture and is spent, but that is not really what happens. It is the suppression of the updraughts that inhibits the cloud and it dissipates not because it loses its moisture but because the cessation of condensation means water is no

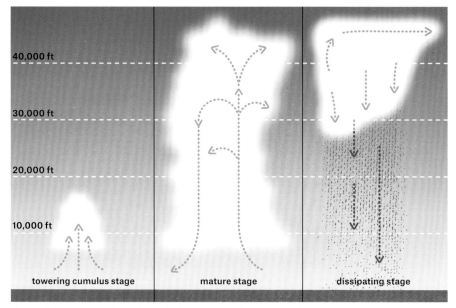

40,000 ft		
30,000 ft		
20,000 ft		
10,000 ft		
towering cumulus stage	mature stage	dissipating stage

A thunderstorm develops through three stages, from a cumulus cloud that is growing into a cumulonimbus, to a mature stage when the cloud produces heavy precipitation and lightning, to a final stage where the cloud dissipates.

longer being added. The illustration above shows the three stages in the life cycle of a storm, the growth stage when the cumulus is towering upward, the mature stage, and the dissipating stage.

Lightning is dangerous and it can strike within 15 kilometres of the storm centre. If you can hear the thunder, therefore, you are within range of the next lightning stroke. Stay out of harm's way.

Squalls and squall lines

A big, black, cumulonimbus storm cloud is a mountain of violent turbulence. It is black because of its size. It towers so high that much of the sunlight passing through it strikes cloud particles and is reflected or scattered before it can reach the eye of an observer on the ground. But not all of its turbulence is contained within the part of the cloud rendered visible by the condensed water vapour. The currents carrying moist air upward produce an area of very low air pressure beneath the cloud, drawing in air. That is how the cloud feeds its updraughts, which originate outside the cloud, close to the ground.

Air is drawn into the base of the cloud in the form of a strong and often gusty wind. If the gusts temporarily increase the wind speed substantially the storm

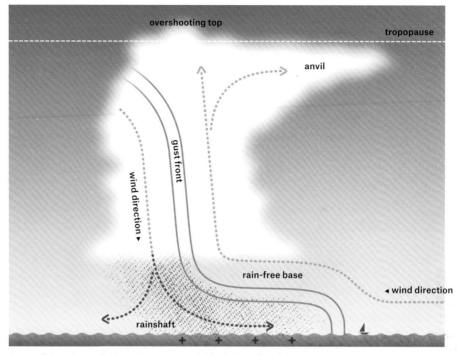

overshooting top

tropopause

anvil

gust front

wind direction ▼

rain-free base

◄ **wind direction**

rainshaft

A squall is a storm that produces powerful but short-lived wind gusts.

is called a squall, defined meteorologically as a wind of at least 30 km/h sustained for at least two minutes, but gusts in a strong squall can reach 100 km/h or more.

The illustration above shows what is happening. Updraughts in the storm cloud are so vigorous they have built the cloud higher than the tropopause, as an overshooting top. Winds just below the tropopause have drawn the top of the cloud into an anvil or incus shape. No precipitation falls from the front of the advancing cloud, where air is entering and rising, but there is heavy precipitation at the rear, in the form of a rainshaft. As the rain leaves the cloud base and enters unsaturated air, some of it evaporates, absorbing latent heat that chills the adjacent air and produces a cold downdraught. Part of the downdraught passes forward, beneath the cloud, and mixes with the air entering the updraught. This produces strong wind gusts along a clearly defined line, the gust front.

Squalls often form along cold fronts, where advancing cold air is pushing beneath warmer, moist air and forcing it to rise. The moist air is conditionally unstable and once it rises above the lifting condensation level it continues upward. There is a weak temperature inversion in the warm air ahead of

A squall line is approaching the shore, producing heavy rain that is clearly visible.

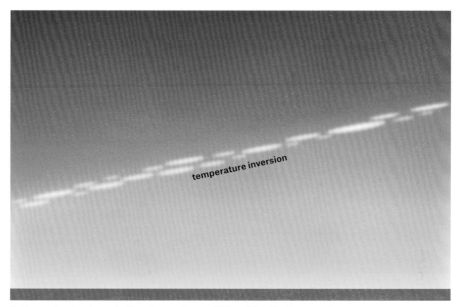

An inversion is a layer in the atmosphere within which the temperature remains constant or increases with increasing altitude.

the front. This is a layer of air in which the temperature either remains constant with increasing height or rises. The illustration above shows the relation between height and temperature in an inversion. Warm air cannot rise through the inversion, because it encounters air at the same or lower density. That confines convection to the frontal surface, where a line of vigorous storms may develop. This is a squall line.

Each storm is short-lived. Ordinary, multi-celled thunderstorms seldom last longer than an hour or two, but along a squall line each individual storm triggers another as it dies. The storm begins to dissipate almost as soon as it reaches its maximum development, but as it was growing its gust front was pushing beneath the moist air ahead of it and to its right, and by the time the storm starts to die down the gust front has raised enough unstable air to start a new cloud forming. This process is so vigorous that the line of storms begins to advance faster than the cold front that gave it birth. It advances ahead of the front as a line of storms up to 1000 kilometres long—a squall line.

The squall line does not remain straight. Changes in wind speed and direction 3–4 kilometres above the ground cause the line to form the shape of an archery bow. This shows in radar images as a bow echo that eventually becomes a comma echo. The illustration on page 148 shows how this appears.

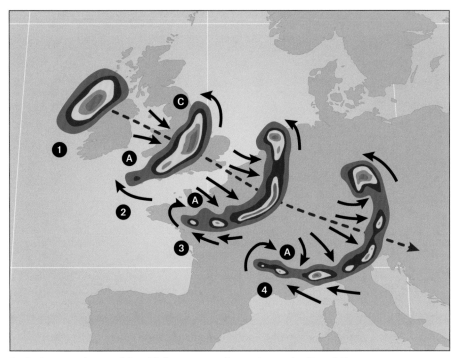

Radar images of a squall line (1) that starts to form a bow echo (2 and 3) as it advances, and finally a comma echo (4). There are regions of cyclonic (C) and anticyclonic (A) airflow. The arrows indicate the wind direction.

At one end of the line, air circulates cyclonically (anticlockwise in the Northern Hemisphere) and at the other anticyclonically. The system is most likely to generate tornadoes at the cyclonic end.

A storm with winds of up to 130 km/h that blow in a straight line, often associated with the gust fronts of a line of thunderstorms, is known as a derecho and can cause severe damage. Derechos can last for up to half an hour and may occur almost anywhere and at any time of year, although they are most common in summer.

Cloudbursts

We often describe a heavy shower as a cloudburst, but a genuine cloudburst is more dramatic. In the course of six hours on the night of 31 July–1 August 1976, more than 300 mm (12 in.) of rain fell on Big Thompson Canyon, Colorado. It sent a wall of water 9 metres high through the canyon, killing more than 130 people. On 29 November 1911, 61 mm (2.4 in.) fell in 5.5 minutes at Port Bell, Panama. These were cloudbursts.

There is no precise definition of a cloudburst, but most meteorologists would use the term to describe rainfall of at least 100 mm (4 in.) per hour over a local area. If you're caught outdoors in rainfall that heavy, the raindrops can hurt, but it can be freakishly local, with a cloudburst on one side of the street and dry ground on the other. If you're driving, the wipers will not be able to cope with the rain and it will bring traffic to a halt.

The name is misleading. At one time people believed clouds were bags, holding their moisture inside a thin, flexible skin, like a party balloon. If the moisture content became too heavy for the cloud, its skin might burst, releasing all that water at once. Clouds are not like that, of course. So what is it that makes a cloud behave in this way?

Obviously, the cloud must be a very large cumulonimbus storm cloud. Already it will be delivering showers, possibly heavy ones. What is needed is a sudden large increase in condensation. That can happen if the cloud approaches high ground and is forced to rise, pushing more warm, moist, extremely unstable air into updraughts that are already rising very vigorously. It can also happen if the cloud drifts over ground that is warmer than its surroundings, such as sunbathed sand or bare rock adjacent to water or moist soil. Warm air is rising strongly from the warm ground and as the cloud passes overhead this additional column of rising air joins it. Again, this accelerates the updraughts and increases the rate of condensation.

Up in the cloud, some of the cloud droplets are merging into raindrops. As these fall the larger of them sweep up the smaller ones, but on an intensifying scale until the streams of coalescing drops grows into a torrent. Then it leaves the cloud as a sudden, huge intensification of the rain. It has become a cloudburst.

In the 1990s, scientists discovered atmospheric rivers, previously unknown phenomena that bring heavy precipitation and may be the cause of most cloudbursts. Atmospheric rivers are streams of moist air typically up to about 300 kilometres wide and 1000–2000 kilometres (sometimes up to 3000) long that transport moisture away from the equator. Usually between three and five of them are active at any time, they survive for up to three days, and they form in the unstable conditions where polar and tropical air meet and form pairs of cold and warm fronts. The moist air rising up the advancing warm front forms a flow called a warm conveyor belt that transports the moisture, and if the air circulation around the frontal system is strong enough the warm conveyor belt can develop into an atmospheric river. A large atmospheric river carries more water than the Amazon and an atmospheric river gathers more moisture as it crosses the ocean. It is when it makes landfall that an atmospheric river delivers precipitation heavy enough to cause serious flooding. Those that cause flooding in California are nicknamed the Pineapple Express, because they originate around Hawaii. About ten of these streams reach Britain every year.

Cloudbursts are short-lived. Atmospheric rivers die away. A cloud rising up a mountainside will reach the summit. A cloud drifting over warm ground will reach past it and move over a cooler surface. The source of the rising air will disappear and the cloud will revert to its normal behavior.

Cloudbursts can happen anywhere, although they are most common in the tropics, where the warmth and abundant moisture allows storm clouds to tower to a great height.

Tornadoes and waterspouts

Its approach sounds like the roar of a freight train, and the monster is clearly visible, hanging beneath its black cloud. It is approaching, but no one can tell whether it will arrive, for its path is erratic and unpredictable. But if it does arrive it will wreak havoc. A tornado is the ultimate expression of the power of a giant storm cloud. They do not occur close to the equator, but they can happen anywhere else and at any time, even at the English seaside.

Late one wet and windy Saturday afternoon in October 2000, a couple were driving through Bognor Regis when their car was engulfed in a downpour of rain and a wall of flying debris appeared no more than nine metres in front of them. Then their car was lifted from the ground, carried sideways a short distance, and dropped on the opposite side of the road, its occupants shaken but unharmed. They had survived a small tornado that ripped chimneys from roofs and hurled roof tiles in all directions. It was a minor event of the kind weather forecasters call mini tornadoes and they are quite common. Britain experiences about 50 every year. Most happen in the countryside and many escape notice. It is when they occur in towns that they make it onto the evening news. And Britain is lucky because its tornadoes are weaklings, with winds of little more than 100 km/h. Those who live on the Great Plains of the United States must face tornadoes with winds of 400 km/h or even more.

Tornadoes are born inside cumulonimbus storm clouds, so they accompany fierce thunderstorms, but these are not ordinary common or garden storm clouds. In most storm clouds there are several convection cells, in which warm air is rising and cold air is descending, dragged downward by raindrops and hailstones. The updraughts and downdraughts are fairly close together and it's not long before the cold downdraughts suppress the warm updraughts. When that happens the cell dies and since this is happening to all the cells

A tornado, the vortex extending from the base of a supercell storm cloud and visible because its low pressure causes water vapour to condense. The cloud around the base consists mainly of dust and debris blown aloft by the air spiralling into the core.

A waterspout is a tornado that either forms over water or that moves over water from the land. Most form over the sea, but close to shore where the water is shallow.

the cloud itself cannot survive for long enough to acquire the power needed to drive a tornado.

There is an alternative, however. If the wind direction at the level of the upper part of the cloud is different from that at lower levels—a condition called wind shear—the updraughts will be directed to the side. The first consequence of this is that they move away from the downdraughts, which are no longer able to interfere with them. This allows a single set of updraughts and downdraughts to dominate the cloud and form a single convection cell, a supercell. The diagram on page 152 shows the structure of a supercell storm cloud. The updraughts are so strong that they overshoot the top of the cloud, sometimes by as much as three kilometres, before subsiding. This produces a distinctive hump on the top of the cloud.

Air rising by convection produces a region of low pressure at the base of the cloud and the updraughts consist of air drawn into this low-pressure region, or cyclone. As the air converges it begins to rotate around a vertical axis. The deflection due to wind shear causes the currents to spiral to one side of the cloud, but if they are strong enough they will lift the entire column of spinning air until it is upright once more. The downdraughts also rotate about their own axes. A downdraught close to the updraughts may then move into an updraught and merge with it. The rotating centre of the cloud is then a mesocyclone. If the downdraught fails to join the updraught it is still possible for a tornado to develop, but it will be weak and unlikely to last for more than ten minutes.

It is also possible for a tornado to develop without a mesocyclone or

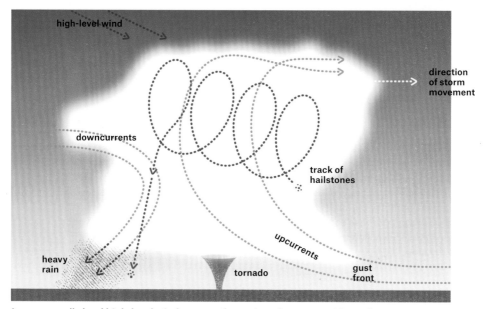

In a supercell cloud high-level winds sweep the updraughts to one side so they are clear of the downdraughts. Since the downdraughts do not suppress the updraughts, the cloud can continue to grow bigger and more powerful.

supercell. These probably begin as swirls of air generated by the air rushing into the updraughts of a growing cloud, but sometimes they form in dry air. Unlike most tornadoes they extend from the ground upward. Most are weak, but there are exceptions. They are called landspouts or land waterspouts. Small tornadoes, called gustnadoes, sometimes develop in the gust front, where the downdraughts spill out from a supercell cloud. Gustnadoes form in air that is diverging and usually spin clockwise in the Northern Hemisphere.

The mesocyclone, occupying most of the interior of the cloud, is now rotating and the low pressure beneath the cloud extends the rotation downward until it projects beneath the rear of the cloud as a section (called a wall cloud) that is turning slowly. Low pressure continues to draw the rotating column of air downward and as the column extends it tapers, forming a funnel beneath the wall cloud. The funnel is just like the vortex that forms when water drains from a bath, but upside down because the flow is upward.

Although air is spiralling upward into the cloud, the mass of air in the column remains constant. The flow simply replaces the air that is leaving the updraughts near the top of the cloud. Any rotating body, including the air in the funnel and mesocyclone, possesses angular momentum. This is the product of

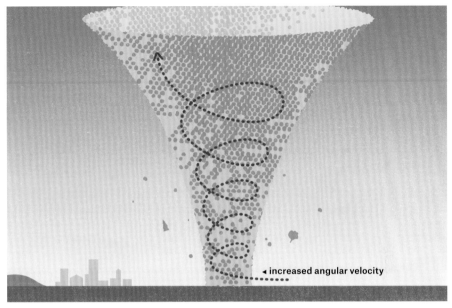

◄ increased angular velocity

A rotating body possesses angular momentum. This is conserved, so as an extending tornado funnel tapers, reducing its radius but without altering the mass of air, the remaining component, the angular velocity, must increase.

the mass (m) of the body, its angular velocity or speed of rotation (V), and its radius of rotation (r), and it is conserved. In other words, mVr is a constant. If one component changes one or both of the others must also change to compensate in order to maintain the constant. In the case of the funnel, tapering reduces the radius. The mass remains constant so the only remaining component, the angular velocity, must increase. The diagram above shows what happens. The conservation of angular momentum explains why the winds are so strong.

The funnel is visible because the pressure in its core is very low. Air entering the funnel expands, which causes its temperature to decrease and its moisture to condense. It is not a section of the cloud somehow drawn downward. It cannot be, because the air is moving upward, into the cloud. If the air beneath the cloud is very dry, the temperature in the funnel may not fall to the dewpoint and water vapour may not condense. In that case, and this does happen, the funnel will be invisible. The funnel becomes a tornado when it touches the ground. As soon as that happens a cloud of dust and debris appears, like a swelling around the funnel's base. The illustration on page 154 shows the structure of a fully formed tornado.

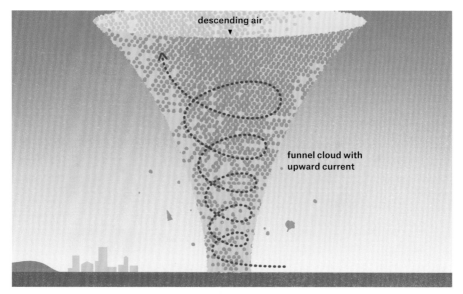

descending air ▼

funnel cloud with upward current

The structure of a tornado, showing subsiding air at the core surrounded by air spiralling upward to form the funnel.

A tornado that moves over water, or that develops over warm water, is a waterspout. Most form over the sea, but they can form over lakes. Most originate in a supercell cloud, but they sometimes develop by convection above warm water, without a mesocyclone. These fair-weather waterspouts are fairly weak, but tornadic waterspouts can be powerful and dangerous. They are visible because of the water vapour condensing in them—they are not columns of seawater drawn upward—and there is a cloud of spray (called a spray ring) around the base where air is flowing into the base of the funnel.

Tornadoes produce freakish effects. They can demolish part of a building but leave the remainder intact, or remove one building without harming its neighbors. Some of these effects are due not to the main tornado but to small tornadoes called suction vortices that may develop around the edges of a large funnel. They are caused by turbulence that produces eddies in the inflowing air and they can be powerful, because they possess the angular velocity of the main tornado plus angular velocity of their own, as well as the forward movement of the main

Tornadoes are very local. Here one has demolished several houses but left one standing in their midst, apparently intact.

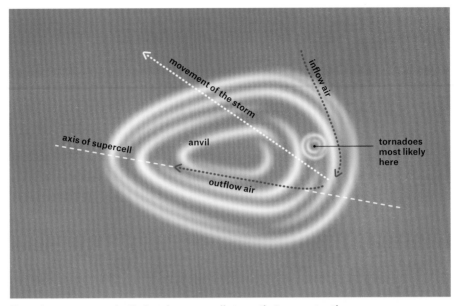

Tornadoes move erratically, but the supercell storm that generates them travels at an angle to its axis. Tornadoes are most likely to appear at the rear of the storm, where air is flowing inward.

tornado, all of which add to their wind speeds. A tornado with winds of 300 km/h can generate suction vortices with speeds of 400 km/h.

Tornadoes move erratically, but the storms that generate them are more predictable. As the illustration above shows, seen from above a supercell storm is approximately egg-shaped and travels at an angle to the right of its long axis. Tornadoes are most likely to emerge at the rear of the storm.

Squall lines often produce tornadoes and the conditions suitable for tornadoes often extend over a large area. Consequently, although tornadoes can occur in isolation, they also occur in swarms, known as outbreaks, and these can be on a large scale, especially in the region of the central United States nicknamed Tornado Alley. On 24–26 May 2011, there was an outbreak of 186 tornadoes, and on 5–6 February 2008, the Super Tuesday Outbreak produced 86 tornadoes affecting 10 states and killing 57 people.

The Operational Enhanced Fujita Scale is the most widely used scheme for classifying tornadoes. Like all such scales its wind speeds are estimates based on surveys of the damage, and they refer to gusts sustained for three seconds. The table on page 156 sets out this scale.

OPERATIONAL ENHANCED FUJITA (EF) SCALE FOR TORNADO DAMAGE

EF number	Wind speed (km/h)
0	105–135
1	136–175
2	176–215
3	216–265
4	266–320
5	More than 320

Hurricanes

A tornado is the most violent of all winds, but it is small, local, and in most cases it lasts no more than a few minutes. A tropical cyclone can be 1000 kilometres in diameter and last several days, sometimes up to two weeks, and although its winds are less violent than those of a tornado, winds in the weakest tropical cyclone exceed 120 km/h, and these are sustained winds, not gusts. So it is violent enough.

Nowadays people usually use the term *hurricane* to describe tropical cyclones regardless of where they occur. Traditionally, though, a hurricane was a tropical cyclone that developed in the North Atlantic or Caribbean, a typhoon one that developed in the Pacific, and a cyclone one that developed in the northern Indian Ocean and Bay of Bengal. No matter what you call them, they all form, grow, and die in the same way.

As their name suggests, hurricanes—tropical cyclones—form only in the tropics. That is because they are fuelled by the latent heat of condensation inside towering cumulonimbus clouds and they require a very warm sea to supply them with a sufficiently high rate of evaporation. The temperature at the sea surface must be at least 24°C (75°F) and hurricanes are most likely to form where the temperature is about 27°C (81°F) over a large area. It is only in the tropics, in fact within latitudes 20° N and S, that the sea is warm enough, and then only in summer.

There is a further constraint. At the equator the magnitude of the Coriolis effect (CorF) is zero and CorF is needed to set the developing storm turning. The CorF is too weak closer than 5° to the equator. Consequently, hurricanes can form only in latitudes 5° to 20°, and even at 5° latitude the CorF is so

Hurricane Katrina in 2005, photographed by a weather satellite. The almost cloud-free eye and spiral structure are clearly visible. Katrina generated winds of up to 280 km/h and caused 1833 deaths in eastern North America.

An approaching hurricane appears as a solid bank of dark cloud.

weak that the storm would have to draw air from a 450 kilometres radius into a cyclone about 30 kilometres across in order to generate winds of about 160 km/h. At latitude 20°, on the other hand, the storm would need to drawn in air from only about 150 kilometres to generate winds of this strength.

Most of the time the tropical atmospheric pressure remains fairly constant, but a small local drop in pressure may be enough to trigger the development of a more persistent tropical disturbance. There are several ways this can happen and if the low pressure drifts westward across the ocean it will gather moisture and clouds will form. The low pressure may fill, but if it survives for several days it may generate thunderstorms and winds of up to 60 km/h. It is then a tropical depression and if it continues to strengthen, when its winds are between 60 km/h and 120 km/h and it shows signs of developing an overall circular shape, it is designated a tropical storm and at this stage it is assigned a name for identification. The storm becomes a hurricane, retaining its name, when its sustained winds exceed 119 km/h and it possesses a distinctive structure.

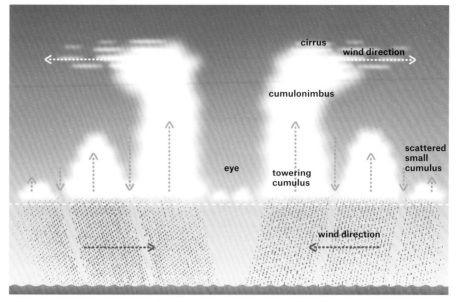

A hurricane comprises concentric circles of convective cloud in which air is rising, separated by bands where air is subsiding and the sky is almost cloud-free, and with a central eye where the air is calm and warm air is subsiding.

The illustration above shows a cross-section through a mature hurricane. Seen from above, the hurricane is circular, with an open centre, the eye. Air is subsiding inside the eye, drawn downward by the low surface air pressure and producing an almost clear sky and warm air that feels still. The warm eye is a defining feature of a hurricane. Technically, when it loses this it is no longer a hurricane. Surrounding the eye there is the eyewall, a solid wall of storm clouds towering beyond the tropopause. These are the clouds that deliver the heaviest rain and strongest winds. Outside the eyewall there is a band of clear, subsiding air, and beyond that a second band of clouds producing rain and wind. This structure, of alternating bands of cloud and clear air, continues to the edge of the storm, 500 kilometres or more from the eye, but with the precipitation and wind decreasing with distance from the eyewall.

The strength of a hurricane depends on the time it spends over the warm ocean. The most violent Atlantic hurricanes begin as storms over Africa that drift out to sea close to Cape Verde, so they are known as Cape Verde storms. They have ample time to develop and if they are fated to grow into hurricanes they achieve that status when they reach the Caribbean or approach the United States coast, travelling at an average 30 km/h.

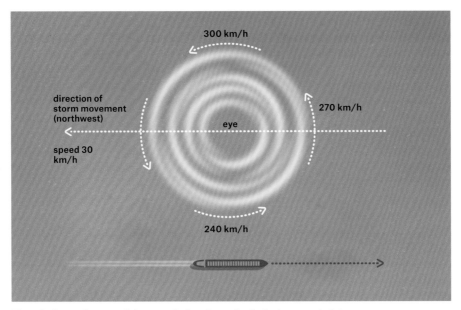

The wind speeds around the eye of a hurricane include the speed of the hurricane itself. Since hurricanes travel in a westerly direction, the wind speed is lowest on the side nearest the equator. This is the safe side to pass. The opposite side is the dangerous semicircle.

Ships receive regularly updated satellite information about hurricanes in their area and are able to avoid them, but in the days before weather satellites, mariners had to rely on their own skills and experience to survive encounters with them and they learned to move to the side of the storm closest to the equator. The diagram above shows why that is what they called the navigable semicircle, contrasting it with the dangerous semicircle on the opposite side. As the hurricane advances, say at 30 km/h, with eyewall winds of, say, 270 km/h, the actual wind speed on either side of the line of the hurricane track contains the hurricane's own speed, making the winds 240 km/h on one side and 300 km/h on the other.

As it continues on its westward journey the Coriolis effect deflects this vast mass of unstable air. It starts to turn away from the equator—to the right in the Northern Hemisphere, to the left in the Southern—and as it does so, moving into higher latitudes, the CorF intensifies, making it turn more until it is moving directly away from the equator. The map opposite showing the track of Hurricane Juliette illustrates this. The map also shows how a hurricane dies as it loses its source of sustenance by crossing dry land. Juliette grew from a

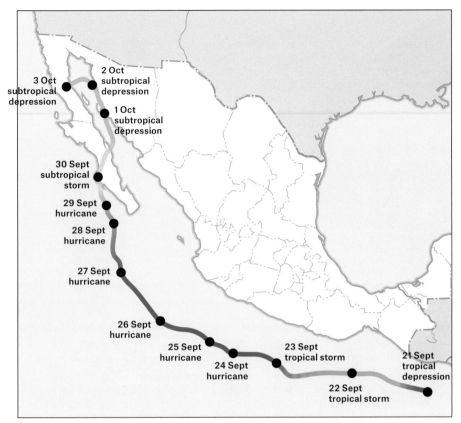

3 Oct
subtropical
depression

2 Oct
subtropical
depression

1 Oct
subtropical
depression

30 Sept
subtropical
storm

29 Sept
hurricane

28 Sept
hurricane

27 Sept
hurricane

26 Sept
hurricane

25 Sept
hurricane

24 Sept
hurricane

23 Sept
tropical storm

21 Sept
tropical
depression

22 Sept
tropical storm

The track of Hurricane Juliette, which briefly reached category 4.

tropical storm to a hurricane on 23 September and reached its maximum strength of category 4 on 25 September. It weakened to a subtropical storm on 30 September when it crossed Baja California and to a subtropical depression the following day, and finally it died down on 3 October. To survive, the storm must remain in contact with a warm sea surface. Hurricanes cannot travel more than a few kilometres over land.

Some Atlantic hurricanes track northward along the US coast, still turning to the right due to the CorF, until they enter the prevailing westerly winds of middle latitudes. These carry them across the ocean, sometimes all the way to Europe. Their track carries them close to the polar front jet stream and the frontal weather systems associated with it and if the dying hurricane encounters a midlatitude depression the two are likely to merge. This will revive the hurricane so it reaches European coasts with renewed vigour and winds capable of causing serious harm. By this time, however, it has long since ceased to

A storm surge occurs when gale-force winds blow toward the shore, pushing water ahead of them. When they strike coastal towns they often cause extensive damage, especially if the surge coincides with a high tide. The debris here was left by a storm surge that struck Sheringham, Norfolk, in December 2013.

be a tropical cyclone. It has lost its warm eye and is an extratropical depression.

Hurricanes are renowned for their winds, but it is the water that does most of the damage. They bring prolonged torrential rain. In 1998 Hurricane Mitch dumped rain at a rate of 305–610 mm (12–24 in.) a day on some places in the Caribbean and Central America. Most feared of all, though, are the storm surges.

Many people assume that because water always finds its own level the sea level is the same everywhere. This is not so, because the sea surface lies beneath

SAFFIR-SIMPSON HURRICANE WIND SCALE

Category	Sustained winds (km/h)	Damage
1	119–153	very dangerous
2	154–177	extremely dangerous
3	178–208	devastating
4	209–251	catastrophic
5	252 or more	catastrophic

the weight of the atmosphere and the sea, unlike the land, can squeeze out of the way of high pressure toward places where the pressure is lower. Beneath the eye of a hurricane, where the air pressure is extremely low, the sea actually bulges upward. The water may rise about 36 centimetres beneath a category 1 hurricane and by 1 metre beneath one of category 5. When the hurricane makes landfall, onshore winds drive huge waves made even larger by the low-pressure "bulge." That is the storm surge and there is yet one more factor that will affect it—the tide. If the storm surge coincides with a high tide it will be huge. In September 1961 Typhoon Muroto II produced a storm surge 4 metres high that sent sea waves surging through the city of Osaka, and in 1992 Tropical Storm Polly caused a 6-metre storm surge at the Chinese port of Tianjin.

Hurricanes are classified by the Saffir-Simpson scale, set out in the table above.

Microclimates

A desert is a dry place where
lost travellers drag themselves through the searing
heat, gasping for water and
struggling toward mirages.
At least, that's the way they
look in B movies. But not all
deserts are dry everywhere.
If there is a depression in
the underlying rocks that is
deep enough for its bottom

If the bottom of a natural depression in a desert is below the water table, it will fill with water and there will be an oasis, a place where conditions are markedly different from those in the surrounding area.

to lie below the water table—for there is water beneath most deserts—then there will be an area of wet ground or even a small lake, with enough water to support plant life. As the illustration above shows, there will be an oasis, and oases can be large, perhaps with gardens to serve as reminders of paradise. The Al-Hasa oasis in Saudi Arabia covers 12,000 hectares and the city of Las Vegas occupies what was once an oasis in the Nevada desert.

An oasis is an area where conditions are markedly different from those in the wider surrounding area but not all are so dramatic or romantic. Urban climates, even those in quite small towns, are different from the surrounding rural climates. On a still smaller scale, the weather is different on two sides of a hill and on either side of a hedge.

Such differences are small but they can be important. All gardeners know that certain plants that thrive in one part of the garden will barely survive in another. The two areas have different climates, or microclimates. Do not despair, then, that your latitude or altitude determine absolutely what you can grow in your garden. There may be a corner with a microclimate that suits plants that otherwise would be exotic, and if there is no such microclimate, perhaps you can make one. A garden pond, a patch of ground deliberately made

A desert oasis develops in a depression where the water table is higher than the ground surface. In this oasis groundwater forms a waterfall where it drops through the roof of a cave. Some oases grow cultivated crops.

A rock garden mimics alpine conditions by producing an appropriate microclimate.

wet, a rockery, or a gravel-covered miniature desert provide microclimates for appropriate plants. So it is time to look at a few microclimates.

Why aspect matters

Study a topographic map of a mountainous region outside the tropics and you may notice that in the Northern Hemisphere most of the villages and towns are located on the south-facing slopes and in the Southern Hemisphere they are on the north-facing slopes. This is for the obvious reason that these settlements and, much more important, their farms and gardens face the equator. Equator-facing slopes are known as adret slopes. They are where most of the food is produced and, therefore, where economic activity is concentrated. Slopes facing the opposite direction are called ubac slopes. These are both names they first acquired in the Swiss Alps.

Adret slopes are warmer because they are more directly exposed to sunshine. They are also drier, because the higher average temperature means the rate of evaporation is also higher. Ubac slopes are cooler and wetter. They are also more likely to be forested, because natural forests have usually been cleared from the valuable land on the adret slopes and because commercial forestry, being less profitable than agriculture and horticulture, tends to be pushed on to the more marginal land of the ubac slopes.

The temperature difference between adret and ubac slopes is sometimes especially pronounced on hillsides that are aligned partly east–west. That is because of the balance between incoming and outgoing radiation. Incoming solar radiation commences at dawn, increases steadily as the Sun rises, and

Adret or equator-facing slopes like this one are warmer and drier than the shaded ubac slopes.

reaches a maximum at noon. Outgoing radiation, however, is proportional to the temperature of the Earth's surface. It is at a minimum at dawn and rises through the morning as the ground warms, reaching its maximum in the middle of the afternoon, after which it declines as the ground starts to cool. Consequently, the ground temperature reaches its maximum not at noon but in the early afternoon. By then the Sun has passed its zenith and is a little way to the west, and shining most directly on slopes facing southwest, or northwest in the Southern Hemisphere. That is why the early afternoon is the warmest part of a fine day.

The east-facing ubac slope receives the morning sunshine but is in at least partial shade by the early afternoon when the temperature on the opposite slope reaches its maximum. Overall, therefore, the east-facing ubac slope is distinctly cooler.

West-facing adret slopes in middle latitudes are also likely to be on the windward side of hills, which means they receive most of the weather systems and orographic rain released when moist air is forced to rise. So they are often wetter than the leeward slopes. Far inland, however, approaching air may have lost much of its moisture and in that case the higher temperature, which reduces the relative humidity, may make the adret slope much drier than the ubac slope.

When planning to build a new home on sloping land, then, it pays to find out the precise direction in which the slope faces. There is more to it than aiming for a south-facing slope, and the aspect will profoundly affect the type of garden you will be able to make and the plants it will support.

How the ground affects moving air

If you plan to measure the wind speed near your home, an anemometer is the instrument you will need. There are several designs but the most familiar is called the spinning cups anemometer. It consists of three small cups that catch the wind and spin about a vertical axis. The instrument shown also has a vane to indicate wind direction, and both are wired to an instrument at ground level where the readings are displayed electronically.

But there's more to it than simply mounting your shiny new anemometer on top of a pole or roof. If its readings are to be reliable the instrument needs to stand at least ten metres above ground level and be sited well clear of buildings, trees, and similar tall objects, because proximity to the surface will slow the wind and tall objects will deflect it.

No ground surface is completely smooth and its small projections resist the movement of air. This is friction and the amount of friction depends on the roughness of the surface. Roughness varies over quite short distances. Think of the effect on moving air that passes in succession over a tarmac drive, mowed lawn, herbaceous border, and tall shrubs. Each of these surfaces is rougher than the preceding one, so each surface, starting with the lawn, presents a leading edge to the airflow. The plants at the edge shelter those behind and friction with the surface slows the wind. The effect is transmitted upward into the air above the plants and the air slows there, but by a smaller amount, and the effect spreads downwind for a distance, called the fetch, that varies according to the roughness and wind speed, until it stabilizes and the airflow has slowed in the layer extending from the ground to the tops of the plants

A spinning cups anemometer, with a vane indicating wind direction. The wire coiled around the arm and stem lead to an electronic display at ground level.

and to a lesser extent in a layer above the plants. The characteristics of the air above this boundary layer are not affected by the vegetation. This process is repeated with each new surface.

Windchill is the sensation we feel when the wind carries away some or all of the warm air stored as a layer in our clothes. We don't experience the opposite effect, of wind warming, because the wind is rarely warmer than our body temperature, but that is not true of the ground surface. Moving air carries away and replaces the layer of air adjacent to the surface, but the efficiency with which it does so depends on the wind speed. The effect is reduced if plants slow the wind, moderating the chilling or warming due to the wind. In cold weather this will keep the surface warmer and in hot weather it will keep it cooler.

A flow of warm, dry air through vegetation also affects the humidity. Imagine the plants in a bed surrounded by warm, dry ground, when a gentle breeze blows along the bed. The wind introduces air that is warmer and drier than the air around the plants near the edge of the border. It raises the temperature and also the rate of evapotranspiration. As the air penetrates deeper into the stand of plants it cools, raising its relative humidity until it is in equilibrium with the air it encounters. This is why in dry weather the soil near the edge of a border is drier than the soil farther from the edge. It is called the clothesline effect, because the wind dries clothes on a line by sweeping away the moist air around them and replacing it with drier air, thus increasing the rate of evaporation.

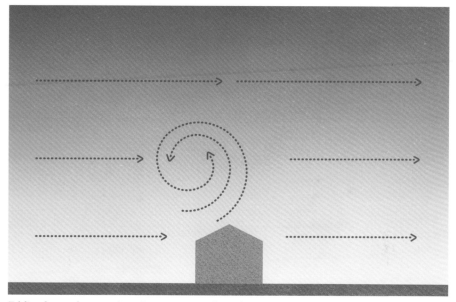
Eddies form when moving air encounters obstacles.

The clothesline effect means that the soil on the upwind side of a bed or garden will be more prone to drying than air on the downwind side and the increased rate of transpiration will expose plants on the upwind side to an increased risk of wilting. An Australian study found that on a mild autumn day with a gentle wind, the clothesline effect raised the evaporation rate by 10% at the edge of a field, and although the soil remained moist, plants wilted in the early afternoon.

If your garden has a fairly large pond or an area that is more or less permanently wet, in warm, dry weather the moisture will alter the climate on the downwind wide. This is called the oasis effect. Water will evaporate into air crossing the pond or wet ground, and the air temperature will fall because the wet surface will be cooler than the surrounding dry ground. Downwind, therefore, conditions will be cooler and moister, and in a dry climate this effect often extends a long distance.

All of this assumes the airflow is laminar; that is, the air flows smoothly, in straight lines, but in the real world roughness causes turbulence, turbulence produces eddies, and eddies produce smaller eddies in a process that continues on an ever smaller scale until it concerns the motion of individual molecules and the natural viscosity of air dampens it. Lewis Fry Richardson, an eminent physicist and mathematician, coined the following lines, known as

As trees deflect and slow the wind, the air swirls around the obstructions, producing eddies that lift dust and small objects, such as these leaves, carrying them spiralling upward.

Richardson's Jingle, to help his students grasp the concept: "Big whirls have little whirls that feed on their velocity, / And little whirls have smaller whirls, and so on to viscosity."

The illustration on page 171 shows a simple pattern of eddies caused by a small building, but in urban areas where there are many buildings, vehicles, street furniture, and other objects to deflect the wind, the pattern is typically complex. When you see little whirls of air that raise dust, dry leaves, and other scraps, these miniature dust devils are caused by eddying. Wind shear also causes eddies at the boundary between two bodies of air moving in different directions or at different speeds.

Eddies make the wind come from all directions and directions that change from one minute to the next. They also mix the air, because they move vertically as well as horizontally, and this can accelerate surface evaporation. Ground is likely to be drier near the corner of a building or other large object, where there are often wind eddies.

Effects of hills and mountains

Mountains occupy almost one-quarter of the Earth's land surface, provide livelihoods for one-tenth of the world's population, and support the world's greatest diversity of vascular plants. The ancestors of our cultivated corn (maize), tomatoes, and potatoes still live in the mountains of Central and South America, and wheat originated in the mountains of the Near East. Mountains may appear scenically attractive but biologically forbidding, but appearances can deceive. Mountains can be highly productive.

There are several reasons why mountains support such high levels of biodiversity. The most obvious is that the climate changes rapidly with elevation. Tropical forests at the base of a mountain give way to coniferous forests at higher elevations, then to grasslands, and finally to a type of tundra, all within a restricted horizontal area. The different climates influence the formation of soil, so soils vary, and variations in gradients determine soil depths. In short, a mountain provides a wide variety of habitats.

Plants and animals that migrate to higher elevations may have requirements that differ from those of the groups that are already established, so they can survive without competing. That is why fears are unfounded that rising temperatures will force species to move to ever-higher elevations until they run out of space. There have been many close observations of what really happens. Species expand their ranges as areas previously too cold for them become habitable, but they do so without expelling the pre-existing species, which adapt readily to warmer conditions. The result is that warming increases rather than reduces biodiversity on mountains.

That said, there is no universally accepted definition of a mountain. In Britain, for instance, any steep-sided hill qualifies if it rises above the surrounding land to more than 600 metres above sea level. The United Nations uses several criteria.

Temperature decreases with increasing elevation at the environmental lapse rate. This varies but the average is 5.5°C (9.9°F) per kilometre, less in moist air and more in dry air. Air pressure also decreases. At 1000 metres it is about 89% of its sea-level value, 78% at 2000 metres, 60% at 3000 metres, and 53% at 5000 metres.

Sunshine becomes more intense with increasing elevation. That is because the air contains fewer solid and liquid particles (aerosols) and less moisture, resulting in less of the incoming solar radiation being absorbed. The extent of the effect varies widely, however, depending on latitude, surface gradient, and aspect.

The overall result is that climbing a mountain is the equivalent of moving

HABITAT ZONES, ALTITUDE AND LATITUDE

Altitude (metres)	0–10°	10–20°	20–30°	30–40°	40–50°	50–60°	60–70°	70–80°
0–1000	T	T	St	St/Tm	Tm	C	C/A	A
1000–2000	St	St	St	St/Tm	Tm	C	C/A	A
2000–3000	Tm	Tm	Tm	Tm/C	Tm/C	C/A	A	A
3000–4000	C	C	C	C	C/A	A	A	A
4000–5000	A	A	A	A	A	A	A	A

T = tropical   C = conifer forest
St = subtropical   A = arctic
Tm = temperate

toward the nearest pole, but with the latitudinal habitat zones greatly compressed. There is an important difference, however. The species inhabiting mountains are more closely related to those at lower levels than they are to species living in higher latitudes. A high mountaintop in the tropics may have an arctic climate, but you will search in vain for polar bears up there. The table above shows how the habitat zones change with latitude and altitude.

Winds on a mountainside are usually different, in speed and direction, from those on the plains. Much of the area is sheltered by other parts of the mountain and less windy than surrounding low ground. Exposed slopes and high ridges, on the other hand, where there is a magnificent view, will experience the full force of a wind that is not slowed by friction with the surface.

Steep-sided valleys may also be windy places either intermittently when the wind blows along them or frequently if they are aligned with the direction of the prevailing wind. As the wind approaches, the high ground bounding the valley constricts the airflow, forcing the wind through the narrow corridor between the valley sides. This is called the funnelling effect, shown in the illustration opposite (above), and high buildings lining a city street also produce it. It accelerates the wind, because air must leave the valley at the downwind end at the same rate as it enters at the upwind end. The wind acceleration will also cause the air pressure to decrease a little near the valley sides.

Mountain valleys also produce their own winds, although you'll feel these only in fine weather, when the air is otherwise fairly still. By day, the sunshine warms the floor and sides of the valley, warming the air adjacent to the surface and causing it to rise up the slope as an anabatic, or upslope, wind. At night the

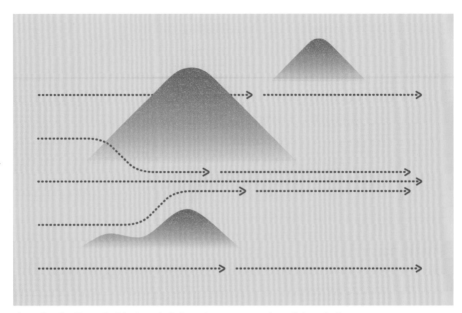

If a valley is aligned with the wind direction, a proportion of the wind will be deflected by the high ground on either side and forced along the valley. This funnelling effect increases the wind strength.

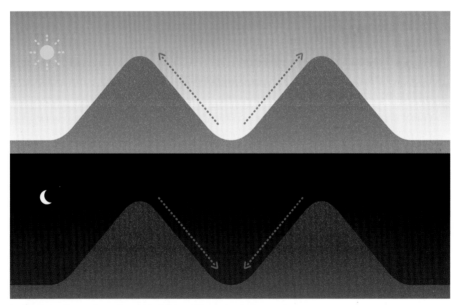

In fine weather, when the air is fairly still, a mountain produces a wind system of its own. By day, air warmed at a low level rises up the mountainside and by night, cool air flows down the mountainside.

ground radiates away the heat it absorbed by day, chilling the air and causing it to sink downhill by gravity beneath the less dense air below. This is a katabatic wind. The illustration on page 175 (bottom) shows the effect.

Precipitation increases with elevation on slopes facing approaching weather systems, but above a certain height, rising air will have lost most of its moisture, so precipitation at high levels is lower than that at low levels. In some cases this produces a high-altitude desert. The lee slopes, in contrast, lie in the rain shadow, and are usually drier. The reduction in precipitation on the upwind side has little effect on plants, however, because the lower temperature also reduces evapotranspiration, so mountain plants utilize water more efficiently than the plants growing on the plain.

Snowline, tree line

Many mountaintops wear a cap of snow, at least in winter. The snowline is the boundary between the snow-covered area and the area that is free from snow. Some mountains remain snow-capped throughout the year and the lowest limit of their snow is also known as the snowline.

The altitude of the snowline depends on the temperature and the temperature depends partly on latitude. You won't be surprised to learn that in the Transantarctic Mountains the snowline is at sea level or that it is rather high at the equator. It is possible, then, to relate the average height of the snowline to latitude. The table opposite sets this out. Note that the snowline is highest not at the equator, but in latitudes 20–30° N and 10–20° S. These are the latitudes where the subsiding air in the Hadley cells produces arid and semi-arid climates. The snowline is higher there for want of precipitation.

You'll also not be surprised to learn that the figures in the table are very general and that the real world is quite a bit more complicated. That is because every mountain has deep gullies that are permanently in shadow and crags that shade the slopes behind them. Slopes that face the equator are usually warmer than slopes that face in the opposite direction, and the snow will fall more heavily on slopes facing upwind than on the lee slopes, in the rain (or snow) shadow. So there is an alternative definition for the climatic snow line. This is the altitude above which snow will accumulate over an extended period on level ground that is exposed to sunshine, wind, and falling snow. At all lower levels ablation will remove each fall of snow before the next fall occurs. The snowline is also higher on mountains that are far from the ocean because they have a drier climate and where snowfalls are light ablation quickly clears them so the snow cannot accumulate.

MEAN SNOWLINE AND LATITUDE

Latitude	N. Hemisphere (metres)	S. Hemisphere (metres)
0–10	4727	5310
10–20	4727	5610
20–30	5310	5125
30–40	4300	3020
40–50	3020	1495
50–60	2010	793
60–70	1007	0
70–80	503	0

On the lee side of many mountains the subsiding, warming air develops into a wind that clears the snow. Winds of this type were first described in the European Alps and are called föhn winds. The North American variety is called the Chinook. They occur in high mountain ranges that lie across the path of the airflow (aligned north–south in middle latitudes). A föhn wind will blow in air that has been forced to rise as it crosses the mountains, losing its moisture, and that subsides on the lee side, warming adiabatically. Alternatively, the wind can also occur when a temperature inversion lies above the mountains. Approaching air is forced to rise but cannot penetrate the inversion, so cloud does not form, but air approaching above the level of the inversion is not barred. It crosses above the inversion then slides down the lee side. Either way, a föhn wind can be strong and can raise the temperature rapidly. The Chinook has been known to raise the temperature by 22°C (40°F) in less than five minutes and to blow at 160 km/h. Any lingering snow doesn't stand a chance. The Chinook is sometimes called the snow eater.

At some distance below the snowline there is the tree line, which is the upper limit for tree growth. This is also the boundary between tundra vegetation and bare rock, and it usually occurs where the average summer temperature remains below 10°C (50°F), which is the lowest temperature most trees can tolerate. On mountains the height of the tree line varies with latitude and with the degree to which the climate is continental or oceanic.

Although the tree line is clearly defined, the approach to it is gradual. It is not like the edge of a forestry plantation. With increasing height, the character of mountain forests, known to ecologists as montane forests, changes. Broad-leaved trees become fewer, coniferous trees more common, but they grow in

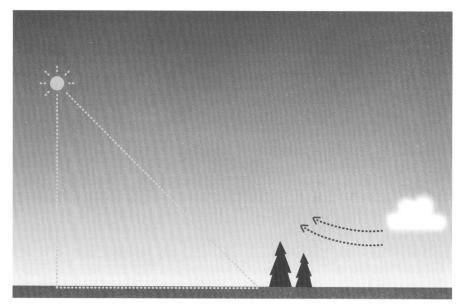

A suntrap is a small area that faces the midday Sun and is sheltered from the wind.

dense stands. Then the trees become more widely spaced until they are scattered in an open landscape. The trees themselves are upright, but at higher levels they are more stunted. The trees close to the tree line are widely scattered, gnarled, stunted, and little taller than shrubs. Between them, herbs grow in soil trapped in sheltered cracks and crevices, there are mosses where moisture collects, and lichens grow on the trees and on the exposed rock.

Suntraps and frost hollows

You wouldn't really want to trap the Sun, but a suntrap is more modest. It is a small area with a microclimate that is warmer and sunnier than its surroundings. What's more, if your garden doesn't already have one it's not too difficult to make one.

As the illustration above shows, an area destined to become a suntrap must face the position of the Sun at the warmest part of the day. This is early in the afternoon, when the ground has warmed and is warming the air, and the Sun has only just passed its zenith. In the Northern Hemisphere, therefore, the site should face approximately south–southwest (about 200°) and in the Southern Hemisphere north–northwest (about 340°). Obviously, the area should be level or on a Sun-facing slope and open to receive the sunshine, but it must be

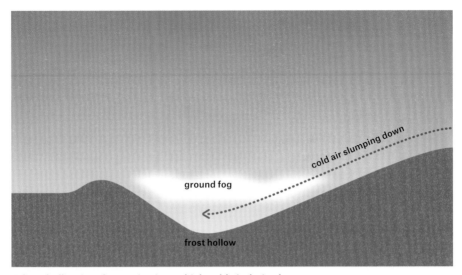

A frost hollow is a depression into which cold air drains by
gravity on clear, still nights.

sheltered from the wind by tall plants or a fence around the sides and rear. In
the illustration, these form an arc, shown by the broken lines, enclosing the protected area. This barrier should not be solid, like a wall, but sufficiently open to
allow air to circulate. A solid barrier will prevent the wind passing, but instead
it will cause air to eddy downward and flow back across the area. Your suntrap
will then trap the wind as well as the sunshine, rather defeating the object.

In fact, of course, it is not so much the sunshine your suntrap is accumulating, but warm air. After all, you could face fully into the sunshine in a howling
gale. The aim is to allow the sunshine to warm the ground and air, and to retain
that warm air for as long as possible.

If there is an opposite to a suntrap it's probably a frost hollow, also known
as a frost pocket, which is a low-lying area or depression that is frosty more
frequently than the surrounding area. The frost forms on clear, still nights
when the ground and objects on the ground, including plants, rapidly radiate
away the warmth they accumulated by day. As their temperature falls they chill
the surface air, which contracts and becomes denser. The dense air then slides
downhill by gravity, pushing beneath warmer, less dense air, and settling into
the hollow or pocket, as the diagram above shows. Toward the end of the night
the air and ground temperature in the hollow may be tens of degrees below that
outside the hollow.

The depth of the cold air is almost equal to the depth of the depression. If moist, warm air then drifts over the top of the cold air a thin layer of advection fog or stratus cloud is likely to form, and the warmer air will form a temperature inversion, trapping the cold air. The frost will be slow to clear because the air cannot mix with warmer air, and if sunshine penetrates to warm the ground it will take some time to raise the temperature sufficiently for the air to be able to rise through the inversion.

It is easy enough to identify a frost hollow so you can choose only frost-tolerant plants to grow there, but the only way to cure the problem involves landscaping the area to expose the hollow to the wind, which will help mix the air, or alternatively to fill in the depression.

Where to expect snowdrifts

In spring, when the snow melts, there are places where it lingers longest, often near the base of walls and hedges. That is partly because walls and hedges that are aligned parallel to the contours of a hillside can trap subsiding cold air, forming frost hollows. But that isn't the main reason snow lingers. It is because these are places where drifts form. By the end of winter it is where the snow lies deepest and the snow is often in the shade of the wall or hedge, so it is slow to melt. But why did the drifts form in the first place?

When snow floats down vertically through still air it settles evenly, and when the snowfall ends the landscape lies beneath a blanket that is of much the same depth everywhere. When there is a wind, however, that is not what happens. The wind carries the air horizontally, and the snow with it. Moving air actually transports the snowflakes, and to do that it must expend energy, just as the happy householders must expend energy to shovel the fallen snow off their driveways. And as all snow shovellers know, snow is heavy. It is water, after all.

The amount of energy moving air possesses is reflected in its speed. The faster the wind blows, the more energy it has, and a strong, energetic wind carries the snow almost horizontally. The more energy the wind has, the more snow it can transport, which is why blizzards are so dangerous. Losing some of its energy will reduce the wind's capacity to transport snow by a proportionate amount.

When you walk into a wind the air presses against you. You can feel it and, if the wind is strong enough, it will make walking difficult or even impossible. In order to push against you, the wind must expend energy, just as you would have to expend energy if you pushed against an object. Imagine pushing a full wheelbarrow uphill. It's hard work and the effort will tire you, leaving you with less energy for the next task.

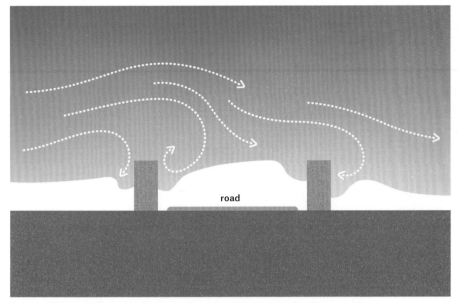

Snow will form drifts on both sides of a wall, with a depression between the wall and the top of the drift. Each of the walls or tall hedges lining a road will produce two drifts and, if the road is narrow, the downwind drift on one side may merge with the upwind drift on the other, blocking the road.

Think about a wall or hedge. The wind pushes against it, expending energy doing so and thereby reducing the energy it possesses. With its energy reduced the wind is less able to transport snow, so it releases some of it and the released snow falls against the obstruction where, if the snow-bearing wind blows from the same direction for long enough, it forms a drift.

Naturally enough, the real world is a bit more complicated. As the illustration above shows, the moving air forms eddies as it encounters the obstruction. Some of the air moves down the obstacle and back and some that crosses the obstacle forms a circular eddy on the lee side. These sets of eddies scoop out some of the snow immediately adjacent to the obstacle on both sides, so in the end there is a drift on each side separated from the obstacle by a small gap.

That also explains the snow's irritating habit of blocking small roads bordered by high walls, hedges, or banks. In this case the barriers on either side of the road collect two drifts each, but the downwind drift behind one barrier merges with the upwind drift in front of the opposite barrier. When two drifts merge in this way the result is a drift of a depth that is the sum of the depths of the individual drifts—and a blocked road.

The banks on either side have produced eddies in the snow-laden wind that have dumped snow until it blocked this small country road.

It is possible to prevent this from happening by using snow fences. These are fences that slow the wind and are designed to accumulate a low drift on the upwind side and a much deeper drift on the downwind side. Erect the fences on both sides of the road at a distance from the road equal to ten times the height of the fences and they will keep the road open.

Stands of vegetation also slow the wind by friction, so they, too, will accumulate snow, but not in the form of drifts. The plants do not present a solid barrier and eddies make the air turbulent, so the snow falls evenly.

We talk rather glibly about the depth of snow. This is misleading because snow is not all the same. The temperature of the air through which they fall determines the size of snowflakes, their size determines the amount of air they will trap between them as they reach the ground, and the amount of air they contain determines the snowfall's volume. It's obviously useful to know the

SNOW TO WATER RATIO

Temperature (°C)	Snow-to-water ratio
Less than −18	40:1
−18 to −13	30:1
−13 to −7	20:1
−7 to −2	15:1
−2 to −1	10:1
2	7:1

depth of the snow before you step outside, but that depth is a very poor indicator of the amount of precipitation. The only way to measure that is to push an open-ended cylinder to the base of the snow, extract the core, and melt it. That will provide the rainfall equivalent of the snowfall, and that is the way meteorologists measure snowfall. As a rule of thumb you can divide the depth of snow by ten to give its rainfall equivalent—a 10:1 ratio—but the table above provides a more accurate measure based on the air temperature.

Hedges, fences, and the wind

If your garden is windy you may wish to protect it by erecting or growing a shelterbelt or windbreak, a barrier placed across the path of the prevailing wind. It sounds simple but if you've tried it you may have found it doesn't necessarily work the way you would wish.

As the air approaches the barrier it is forced to rise and the air is squeezed at the top. This accelerates the wind, but as soon as it is on the lee side the airflow widens and the wind decelerates. As it crosses, the air forms large eddies on both sides of the barrier. On the lee side these eddies continue for a distance downwind before the airflow becomes smooth once more and the wind recovers its original speed. The diagram on page 184 shows the effect. A barrier will affect the airflow above and on the upwind side by a distance equal to about three times the height of the barrier.

What happens downwind depends on the barrier itself. If the barrier is solid or very dense, like a wall or thick hedge, all or most of the air will have to go over the barrier and the wind speed will be greatly reduced in the large eddy on the downwind side. It will quickly recover about 90% of its speed, however, so the barrier will be effective for a distance equal to only about 10–15 times its height. If the barrier allows some of the air to pass through, no downwind eddy

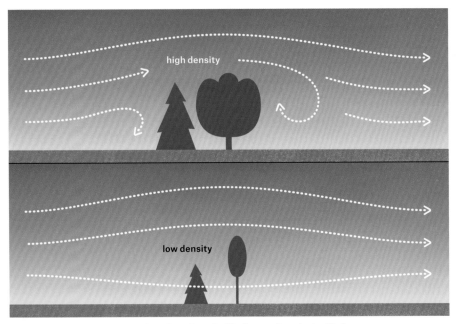

If a barrier planted or erected to slow the wind is dense the wind will slow down, but form eddies on either side and soon recover its former speed. If the barrier is of low density, allowing wind to move through as well as over it, the wind is slowed less but the slowing extends farther downwind.

will form. The reduction in wind speed will be less, but the effect will extend to a distance equal to 15–20 times the height of the barrier. It is possible to improve on this with a barrier of medium density, which allows just enough air to penetrate to prevent eddying on the lee side. It can reduce the wind speed to below 90% of its upwind value for a distance equal to 20–25 times the height of the barrier and there are some shelterbelts that have a downwind effect that extends to 40 times their height.

Town climate, rural climate

Engine exhausts from trains and road vehicles discharge warm air. Factories that release almost no pollutants do release warmth. In summer, air conditioning systems discharge into the streets and in winter, heated buildings leak warm air. And on top of it all, the body warmth of thousands or even millions of people, domesticated animals, and the wild birds and mammals that seek our company combine to make cities warmer than the countryside beyond the city limits. It is called the heat island effect and it is considerable. By late evening a

Los Angeles, CA, smog results from photochemical reactions involving vehicle exhausts.

city centre is commonly 5–6°C (9–11°F) warmer than the adjacent countryside and it is sometimes as much as 8°C warmer, with the temperature falling fairly steadily as you move away from the centre and out through the suburbs.

Even without engines and bodies to generate heat, the city itself would still be warmer than the countryside. Concrete, asphalt, and bricks have a lower albedo than farm fields. They absorb up to six times more heat and release it rapidly after nightfall to warm the evening air, reducing the diurnal temperature range. This can be beneficial, because large differences between day and night temperatures are harmful to health.

City buildings also reduce the wind speed, so cities are 25% less windy than the countryside, another factor in making them warmer, and there is an additional effect. The larger the city, the slower the wind is by the time it reaches the centre, but this reduces the extent to which air is mixed with outside air. Consequently, the air that is warmed by all the activity around the city centre remains there for much longer than a comparable volume of air in the open countryside, which would quickly be mixed with cooler air.

When the wind does blow, however, cities with long, straight streets may suffer from the funnelling effect. Tall buildings on either side make a city street resemble a canyon. The buildings will funnel and accelerate the wind along streets that are approximately aligned with the wind direction. That's when the umbrellas take on a life of their own.

Less paradoxically than it may seem, the calmer city air sometimes combines with the temperature difference between urban and rural areas to produce a breeze of its own, by a mechanism similar to that which generates land and sea breezes, but usually in only one direction, as a country breeze. Warm city air expands, producing an area of low atmospheric pressure, which draws in cooler air from outside. The country breeze is strongest in the afternoon, when the temperature difference is greatest, but it often continues through the night, because although the city cools at night its temperature seldom falls to that in the countryside.

City air is also drier. Rain falling on streets and roofs and snow melting from them cannot soak into the ground as it would into soil. Some evaporates, but most surface water is channelled into storm drains that remove it. If it were left to dry by evaporation the streets would be wetter than they are and also cooler, because evaporation would absorb latent heat. Nor are there so many plants in the city, even allowing for the parks and gardens, and that means there is less transpiration to return moisture to the air. The relative humidity is an average 6% lower in the city.

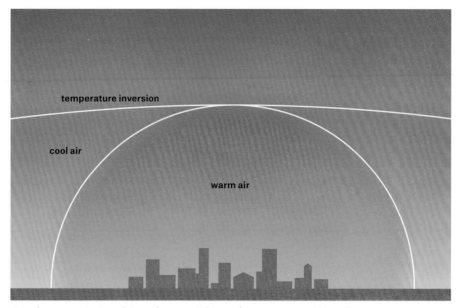

temperature inversion

cool air

warm air

An urban area sits in a dome of warm air beneath a temperature inversion.

Strangely, although urban air is drier than country air, the city receives 5–15% more precipitation than the nearest rural area, and its skies are 5–10% cloudier. No one is quite sure why this should be so, but some factors are known. City air is dustier than country air and the dust provides higher concentrations of cloud condensation nuclei, favouring cloud formation.

Warming the air close to ground level also tends to make urban air unstable. This favours the development of cumulus clouds, some of which grow into cumulonimbus storm clouds. City dwellers experience about 5% more winter thunderstorms than people living in rural areas, and almost 30% more in summer.

On the whole and despite the rain and storms, this makes it sound as though the warmer, less windy weather in cities is rather more pleasant than country weather, but there are downsides. The dusty air reduces the intensity of the sunlight and the buildings shade many areas, so 15–30% less sunshine penetrates to street level than in the open countryside, and ultraviolet radiation is 5% less in summer and 30% less in winter.

Urban air quality is also poorer, especially in winter. In winter, rising warm air often produces a temperature inversion above the city, a layer of air in which the temperature increases with height. Rising air is unable to penetrate

London pea-souper fogs were once notorious, and very harmful to health. They were common in Victorian times, as see here, and remained a regular winter occurrence until the 1950s, when legislation banning the burning of coal brought them to an end. Similar winter fogs occurred in many other industrial cities.

the inversion and it spreads to the sides. As it does so, the air radiates away some of its own warmth, becoming denser as its temperature falls, and subsides over the surrounding countryside. This produces a region of low pressure around the urban centre and high pressure outside, so air flows back into the city close to ground level, is warmed and rises, and so a convection cell becomes established beneath the inversion, with a domed shape, as shown in the illustration on page 187. The circulating air remains confined within this urban dome and, consequently, any pollutants it carries accumulate. The notorious pea-souper smogs that were once a regular winter feature in most industrial cities, especially London, occurred when moist air mixed with smoke from coal fires became trapped beneath an urban dome. Legislation

MICROCLIMATES

mandating the use of smokeless fuels ended those smogs, but other pollutants continue to cause problems in many cities when they accumulate in this way.

Below the dome the climate is strongly influenced by the many microclimates produced by the buildings, streets, parks, and gardens at or close to ground level. The roofs of the buildings then resemble the canopy of a forest, and it is called an urban canopy, the air beneath it forming the urban canopy layer.

Urban heat islands occur even in quite small towns. Barrow, Alaska, at latitude 71.3° N and with a population of about 4500, has one. Studies over many years have found temperatures in Barrow are an average 2°C (4°F) warmer than those in the surrounding rural area, sometimes rising to 4°C (7°F) or even up to 8°C (14°F) warmer. And the effect has been known for a long time. The first person to describe it was Luke Howard, the English chemist and meteorologist who devised the system we still use to classify clouds, in *The Climate of London*, published in 1818–19.

Soils

Our concentration on the importance of climate and weather might make it seem that a glasshouse inside which you could control the temperature, humidity, and hours of daylight is all it would take to be able to grow fruit and vegetables in the dry valleys of Antarctica or on Baffin Island. But there is another necessary ingredient—soil. Of course, you could import soil, but then you would have created a complete temperate-climate environment isolated from the world around it. What you could not do is to use the soil available locally, because there is none.

Soil and climate are intimately linked. Examine a sample of soil closely and you will find it composed of mineral particles derived from the underlying rock mixed with organic material derived from the decomposition of plant and animal detritus and water. The processes that convert bare rock into soil, function only under certain conditions of temperature and humidity. In other words, soil formation is driven by climate. Depending on the climate it may take place rapidly, slowly, or not at all. So it is time to think about soils.

How soils form and age

In the beginning, when volcanic action or movements of the Earth's crust expose a new land surface, there is only bare rock. Between 1963 and 1967 a series of volcanic eruptions on the seabed 32 kilometres south of Iceland thrust a new island above the surface. Icelanders called it Surtsey and today it is a World Heritage Site where biologists monitor the arrival and establishment of living organisms. Its 141 hectares currently support 60 species of vascular plants, 75 species of bryophytes, 71 of lichens, and 24 of fungi, as well as 335 species of invertebrates, and 89 species of birds have visited it. No people live there, so it is a kind of living laboratory where scientists can observe the natural compilation of a habitable environment, based on the production of soil.

Bare rock expands in the summer sunshine and contracts in the cold of winter. Expansion and contraction produce cracks and the minerals in rock expand and contract by different amounts, causing additional fractures. Warm sunshine causes the rock surface to expand more than the underlying rock, detaching the surface layer, which gradually flakes away. This is called exfoliation or onion-skin weathering. Rain fills the cracks and in high latitudes it freezes in winter. Water expands as it freezes with enough force to shatter rock, breaking away particles that wash out when the ice melts. Wind and rain detach and remove fragments of rock and roll and grind loose rocks against each other and against the solid rock. The rock wears away, loose fragments accumulate, and these, too, gradually become rounded and smaller until huge boulders are reduced to tiny grains. The process is called mechanical weathering. And it is not all.

The crevices are small, but large enough for bacteria, carried in the air and falling randomly everywhere, to find shelter from wind and shade from direct

Where a landslip or cutting exposes a cross-section through the soil the root systems of plants are often displayed.

sunlight. Rainwater also penetrates and natural rainwater is weak carbonic acid because it contains dissolved carbon dioxide. The acid reacts with minerals in the rock, slowly dissolving parts of the rock and becoming a solution that contains substances on which the bacteria can subsist. This is chemical weathering.

The colonies of bacteria secrete their own protection in the form of a jelly-like coating, and as the rock crevices deepen and widen the bacteria expand into them. Bacterial wastes also accumulate and react chemically with the rock. After a time, layers of organic material develop. They are thin, but provide purchase for lichens. With more organic material, mosses are able to establish themselves and some time later the first vascular plants appear. Plants have roots, which penetrate the crevices in search of moisture and nutrients, and in doing so they excrete substances that feed microorganisms, accelerating chemical weathering.

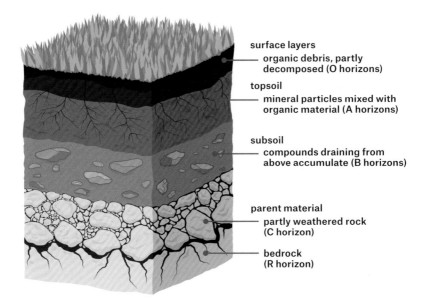

surface layers
— organic debris, partly
decomposed (O horizons)

topsoil
— mineral particles mixed with
organic material (A horizons)

subsoil
— compounds draining from
above accumulate (B horizons)

parent material
— partly weathered rock
(C horizon)

bedrock
(R horizon)

A profile is a vertical section cut through a soil from the surface to the underlying bedrock. It reveals a number of layers, or horizons, each with its own characteristics. The horizons are labelled. This diagram shows all the possible horizons. Not all soils have all of these.

A surface layer of mineral particles is called a regolith. When organic material becomes mixed with the mineral particles, the mixture becomes soil.

All soil is derived from rock and since the process of its formation has a beginning, soil must have a life story. The soil must age and eventually become very old, and the rate at which this happens depends on the climate. In high latitudes, where winters are long and cold and summers short and cool, the process is very slow. In deserts, where aridity inhibits organic activity, the process may never begin and the regolith remains regolith. Near the equator, in contrast, high temperatures and abundant rainfall accelerate the process greatly. So the age of a soil has nothing to do with the time that has elapsed since it first formed, but describes only its stage of development.

Larger plants arrive, with roots that penetrate deeper and allow rainwater—weak carbonic acid—to percolate further into the bedrock, extending the weathering process downward. Dead leaves, fallen twigs and later branches, the remains of annual plants, animal waste, and dead animals form a surface layer that provides food for a large and diverse population of organisms and, little by little, the soil grows deeper. The nutrients the organisms remove from the soil for their sustenance return to the soil through the processes of decomposition, so nutrients are recycled in the soil, though not completely.

As rainwater drains through the soil it removes materials in suspension from the upper layers and transports them to lower layers. This process is called eluviation and it accompanies the process of illuviation, in which compounds are deposited in layers of soil either from above or by being washed in laterally. Weathering continues to release compounds that enter the soil solution and these also move away, to lower layers or adjacent soils. That process is called leaching.

In time, the vertical movement of materials produces distinct layers, called soil horizons. These are illustrated in the diagram opposite of a soil profile, which is a vertical section cut through the soil from the surface to the underlying bedrock. The bedrock supplies the parent material from which the mineral component of the soil is derived. There are many horizons in a fully matured soil, although few soils exhibit all of them, grouped into four principal categories labelled by letters. The O horizons consist of organic debris, some of it partly decomposed. The A horizons are the topsoil, made from mineral particles mixed with organic material that has moved downward by eluviation and leaching. The B horizons form the subsoil in which compounds draining from above accumulate. The parent material, of partly weathered rock, forms the C horizon, and the underlying bedrock forms the R horizon.

At first the young soil has only a thin covering of organic detritus above undifferentiated mineral particles, comprising horizons O and C, but as more plants become established and animals arrive to feed on them, the soil deepens and matures, with the development of more horizons. All the time, though, the soil is losing compounds by leaching and eluviation and as the soil starts to show signs of ageing, the contrast between horizons A and B becomes extreme, as the A horizon is depleted of substances that accumulate in the B horizon. Plants then begin to suffer as the soil becomes less fertile, and with further ageing the fertility continues to decline. The more aggressive plant species that were previously dominant begin to fail as the rich supply of nutrients they demand dwindles. That allows the vegetation to become more diverse as species with more modest requirements flourish in the old soil. Finally, only the least soluble nutrients remain accessible to plants, and the only plants are those with shallow roots that live on nutrients they obtain from recycling organic matter. The soil is then senile and continues to deteriorate until at last it is fully weathered. All its nutrients are then gone and its fertility is extremely low.

Clay consists of flat sheets of minerals present in a jumble (left) and bound together by salts. When the salts dissolve and drain out, the sheets pack together (right) into a dense, relatively impermeable layer.

Types of soil

Every soil has a history, so it is not surprising that the factors shaping them produce a wide range of soils. The first difference arises from the relative sizes of their mineral particles. This really is simple, because there are only three relevant sizes. Soils consist of sand, silt, and clay particles in varying proportions. Sand particles are 0.02–2 millimetres in size, silt particles are 2–20 micrometres (μm), and clay particles are less than 2 μm. Sand and silt differ only in size. Both are made mainly from hard minerals such as quartz and have undergone no chemical weathering.

Clay has undergone some chemical weathering. Its particles consist of alternating layers of silica (silicon dioxide) and aluminium oxide that form flat sheets. Variations in the proportions of silica to aluminium sheets produce a range of clay minerals with somewhat different properties. The particles may be present in a jumble, bound together by salts, but when the salts dissolve and wash away the sheets align to form a dense layer of impermeable material, as

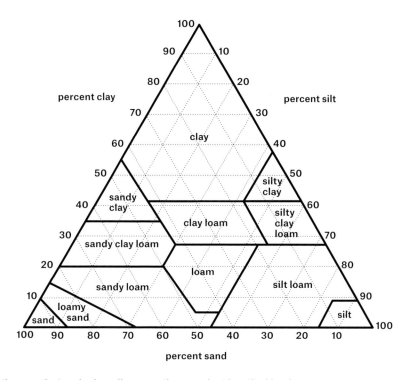

A soil textural triangle that allows a soil type to be identified by the proportions of sand, silt, and clay it contains.

shown in the illustration opposite. That is why clay soil is so heavy and why it becomes so sticky when it is wet.

All soils are based on a mixture of these three types of particles, and to describe them soil scientists have devised a diagram called a textural triangle (above). With sand, clay, and silt at the three corners, the percentage of each determines whether a soil is a sand, sandy loam, loam, silt loam, silt, sandy clay loam, clay loam, silty clay loam, sandy clay, silty clay, or clay.

These names describe the texture of a soil, but they say nothing about its other characteristics, which are determined by the parent material that supplies its minerals, the climatic conditions in which it develops, and its history. The ingredients of soil and soils themselves migrate vertically and horizontally and, as every farmer and most gardeners know, the type and quality of the soil can vary substantially on a quite local scale, resulting in markedly different profiles. There are many different types of soil, many systems for classifying them, and each type has its own name in each of the classifications. This makes naming soils very confusing and the confusion is not helped by the

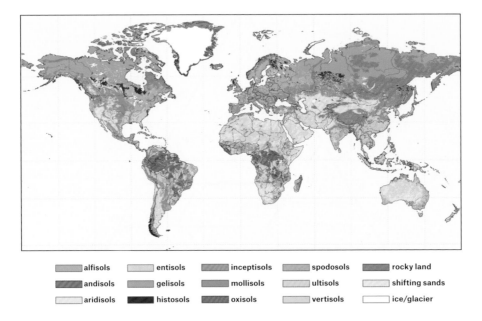

alfisols		entisols		inceptisols		spodosols		rocky land
andisols		gelisols		mollisols		ultisols		shifting sands
aridisols		histosols		oxisols		vertisols		ice/glacier

The global distribution of the 12 orders in the US Soil Taxonomy.

strange nomenclature soil taxonomists have devised. Anyone seeking to pursue this study must become familiar with names such as leptosols, vertisols, ferralsols, gelisols, oxisols, ultisols, and many more.

Most countries have a classification that suits the soils found within their borders. Classifications are usually based on a particular soil horizon that is deemed to be diagnostic because it contains a suite of characteristics typical of that type of soil. The classification devised by the Soil Survey for England and Wales is one of the national systems, and among the most straightforward, with only seven of what it designates soil groups.

Gley soils are often waterlogged for prolonged periods. This leads to the chemical reduction of iron compounds, which then move downward. Their horizons are often grey in colour and often with rust-red mottling. The process of acquiring these features is known in Britain as gleying and in the United States as gleyzation.

Lithomorphic soils are very shallow, with an organic horizon lying above bedrock.

Brown soils are well drained with no sign of gleying in the uppermost 40 centimetres, but with differences in mottling and lessivage—the downward movement of soil particles, especially clay—below 40 centimetres forming the basis for subdivisions. These are the most fertile soils, highly prized for agriculture and horticulture.

Pelosols are clay soils that crack when dry.

Podzolic soils form where rainfall is abundant and the mineral particles migrate readily. The soil is usually sandy and acid, often with a rust-coloured B horizon rich in iron oxides and aluminium oxides.

Man-made soils result from human activities and include soils resulting from the reclamation of mine and quarry wastes.

Peat soils are rich in organic matter in a layer at least 40 centimetres thick.

The United States Department of Agriculture and the National Cooperative Soil Survey have developed a classification known as the US Soil Taxonomy that covers the world and is widely used. It divides soils into 12 orders and each of the orders into suborders, great groups, subgroups, families, and series. The map opposite shows the global distribution of the 12 orders of the US Soil Taxonomy.

There is also an international classification developed by the Food and Agriculture Organization of the United Nations (FAO), the UN Environment Programme (UNEP), UN Educational, Scientific and Cultural Organization (UNESCO), and the International Society of Soil Science (ISSS). It is called the World Reference Base for Soil Resources (WRB) and is meant to be applicable throughout the world, but not to supersede national classifications. The WRB is based on 30 reference soil groups derived from 40 diagnostic horizons.

What matters, of course, is the suitability of a soil for growing plants—its ability to store and release plant nutrients. A complex hierarchy of soil organisms combines to recycle nutrients through the decomposition of plant and animal material, but the initial stock of nutrients is derived either from the parent material through chemical weathering, or by being transported from adjacent land. Nutrients exist as soil colloids or colloidal particles. These are mineral or humus particles that are microscopically small, but much larger than a molecule. They have a very large surface area in relation to their volume and carry a permanent or variable negative electrostatic charge. This attracts and holds cations—ions with positive charge—some close to the colloidal particle, others farther from the colloid surface, and the cations are able to exchange between these locations and between colloids and the soil solution. It is a soil's capacity for cation exchange, its cation exchange capacity (CEC), which determines its ability to store plant nutrients. Soils rich in clay and organic matter have high CEC and are very fertile. Sandy soils and soils with little organic matter have a lower CEC and are less fertile.

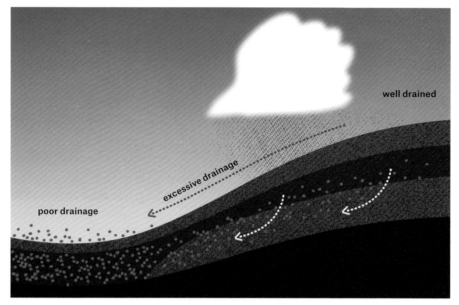

well drained

excessive drainage

poor drainage

Changes in the rate at which groundwater flows is one of the factors that can alter the characteristics of a soil over a horizontal distance. In this example, good drainage at the top of the hill washes mineral particles down the slope, drainage accelerates where the gradient is steepest, and particles and other sediment accumulate near the bottom of the slope.

How water moves through soil

Several things can happen to the water when it rains. If the rain is heavy and falls on fine-textured bare soil it can batter the surface particles, packing them tightly together to form a thin but impermeable skin, called a cap. If the land slopes, the falling water then flows downhill across the surface as surface run-off, and if the surface is level, the water accumulates in pools. If the soil texture is coarser or the rainfall less intense, the water infiltrates, sinking slowly downward, but as the soil becomes wetter the sinking becomes percolation, which is slower and involves descending water displacing water below. After the rain has stopped, any water lying on the surface quickly disappears, some by evaporating but most by draining downward.

At the base of a soil profile there is the bedrock from which weathering releases the parent material. Bedrock is impermeable and above it there may be a layer of clay or other equally impermeable material. The water draining downward through the soil is unable to penetrate this material, so it accumulates above it, not as a subterranean lake, but as saturated soil. This is groundwater and it flows downhill across the surface of the impermeable layer. If the impermeable layer is level or slopes into a depression, the groundwater may reach the surface.

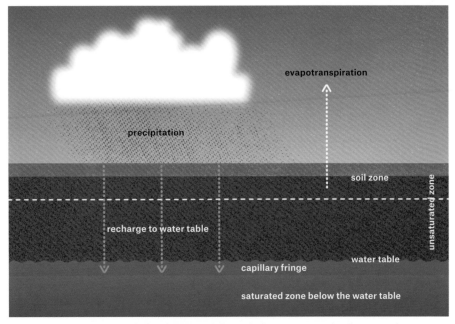

Water from precipitation drains downward through the unsaturated soil zone and joins the groundwater in the saturated zone by recharging the water table. At the same time, some of the water returns to the atmosphere through evapotranspiration. In the capillary fringe, water is moving a short distance upward by capillarity.

Groundwater, headed ultimately to the sea or a lake, moves slowly, typically at a rate of about 2 cm/h, the actual rate depending on the gradient and the permeability of the material through which it is moving. In clay or shale the groundwater may travel no more than 30 centimetres in a century, whereas in limestone with many caves it may cover 1.5 kilometres in an hour.

Gradients and soil material can change over quite short distances, of course, and such changes affect the rate of groundwater flow in ways that produce marked differences in adjacent soils all derived from the same parent material. Such a sequence of soils is known as a catena or toposequence and the illustration opposite shows a typical example on a hillside. At the top of the hill the ground is fairly level and well drained. Water moves downward and starts to flow downhill, carrying mineral particles with it. Farther down the hillside the gradient steepens and water drains much faster, transporting more soil material. The transported material finally accumulates at the bottom of the hill, where the ground is level once more. After a time, the soils at the top, middle, and bottom will be markedly different.

Water supplied by rain and melting snow drain downward through the unsaturated soil. As it does so, some of the water evaporates from the soil

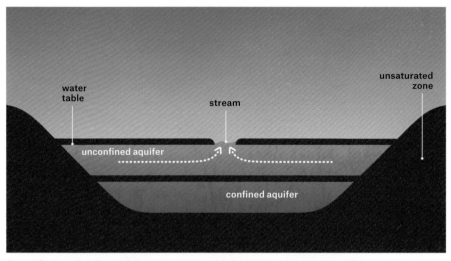

Groundwater that flows above an impermeable layer constitutes an aquifer, and flows as a stream where the water table is higher than the ground surface. If there is no layer of impermeable material above the aquifer it is said to be unconfined, and it is confined if it lies beneath an impermeable layer.

and some enters plants and returns to the air by transpiration. The remainder joins the groundwater, the upper margin of which is the water table. The name makes it sound as though the groundwater has a surface like a lake or the sea, but the groundwater occupies all the tiny spaces between the particles of saturated soil, and rather than being a surface, the water table is a region in which the soil is not quite saturated. The diagram on page 201 shows the overall structure.

Immediately above the water table there is the capillary fringe, a layer in which water is moving a short distance upward by a process called capillarity, or capillary attraction, and evaporating as it enters air spaces in the overlying unsaturated soil. Capillarity results from two properties of water, adherence and coherence. Place water in a container—or in the small spaces between soil particles—and because water molecules carry an electrostatic charge they will cling to the sides and the electrical attraction will make them move along the sides. That is adherence. Coherence is the property that makes them cling to each other, so as water advances along the sides of its container it drags more water behind it. That raises the surface, allowing the water to advance a little farther along the sides, and the movement continues until the weight of water in the column equals the force of adherence. Since a narrow container holds less water than a wide container and, therefore, less weight, capillarity is significant only in very narrow spaces.

SOILS

A spring appears where the water table intersects the ground surface.

Capillarity can also be explained another way. Imagine a basin of water with a narrow tube rising vertically from it. At the bottom of the basin the water is subject to the pressure of the water above it. That pressure decreases with height in the basin, because the amount of overlying water decreases, until it is zero at the surface. Inside the tube, however, the pressure is less than zero because the tube is above the surface, and it is that negative pressure which causes water to rise. For convenience, soil scientists remove the minus sign, and instead of calling the force negative pressure they call it soil moisture tension.

The permeable material through which groundwater flows is known as an aquifer and a well sunk into it will refill as long as its bottom remains below the level of the water table. If the aquifer penetrates the ground surface the water will emerge as a spring or seep that may feed a stream. As the illustration opposite shows, an aquifer lies above a layer of impermeable material. If there is no impermeable material above it, the aquifer is unconfined. A confined aquifer is one held between two impermeable layers.

All wells are valuable, but most need pumping, or a bucket on the end of a rope, to bring the water to the surface. But there is one type that flows freely, without assistance, and it releases water from a confined aquifer beneath a depression in the land surface. Roman engineers were the first to recognize wells of this sort, in the region of Artois in northeastern France, and they are called artesian wells. The water table in the confined aquifer is higher than the ground surface in the depression. Consequently, gravity exerts pressure on the water proportional to the hydraulic head—the height of the water table above a specified level. Drilling through the upper impermeable layer releases the pressure and water flows not to the surface, but to the level of the water table. Since this is higher than the surface, the water is likely to spout forth under pressure. The diagram on page 204 shows a cross-section through land with an artesian well.

Porosity and permeability

Soil contains decomposing organic material and humus, but it consists mainly of mineral particles and it is these that determine how easily the soil retains water. Clay holds water well, sand much less well, and water flows straight through gravel. The difference lies in the average size of the particles.

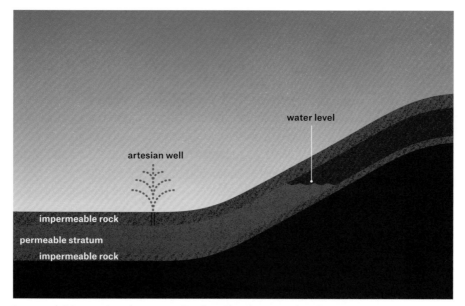

In a natural depression above a confined aquifer, the ground surface may lie below the level of the water table. Gravity then exerts pressure on the water in the aquifer. If a well is sunk into the aquifer, water will rise to the level of the water table without pumping. Since the water table is higher than the ground surface, the water may emerge under pressure.

Apart from clay particles, which form layered stacks, soil particles have irregular shapes and when they pack together there are small spaces (pores) between them. The porosity of a soil is the amount of pore space a sample contains, usually measured as the percentage of the volume of the sample. The table lists the porosity of a few types of soil and rock.

When water moves through a soil it passes through the pore spaces. It might seem obvious that the larger the particles are, the more pore space there will be and, therefore, the more easily water will pass. But it is not necessarily so. In a box with a volume of 170 cm^3 containing 10 spheres each with a radius of 2 centimetres, the total volume of all the spheres will be 33.5 cm^3. Now imagine the same box with 100 spheres each 0.2 centimetres across, and their total volume is also 33.5 cm^3. Both have a porosity of approximately 20%. So what matters is not the total porosity, but the sizes of the individual pore

POROSITY

Material	Porosity (%)
average soil	55
clay	50
sand	25
gravel	20
limestone	10
sandstone	10

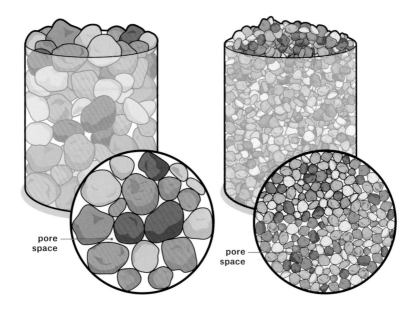

pore
space

pore
space

Regardless of the size of the particles, if the total volume of the large ones
is the same as that of the small ones, the amount of pore space will be
equal in both containers.

spaces. Big particles have larger spaces between them than do small particles, and that is why water passes more easily through gravel than through clay. The diagram above shows two such boxes, one containing many more particles than the other, but both with the same porosity. In a real soil, water would be able to pass between particles, but with that allowance it is easy to see that water will pass more freely among the larger particles.

In ordinary conversation we use words like gravel and sand rather loosely, but in soil science they have precise meanings. There are systems for classifying particles by their size, including British, US, and international schemes. The US scheme, often called the Udden-Wentworth classification, after the two scientists who devised it, has also been modified to a phi (Φ) scale, which gives a simple whole number. The first table on page 206 gives these three classifications and the second table lists the international classification, which works slightly differently.

What matters is not the porosity of the soil but its permeability, which is the ease with which water moves through it. Usually, the finer the soil the less permeable it is. Permeability is measured as the rate at which water moves. The first table on page 207 lists the classes of soil permeability with their rates of water movement.

PARTICLE SIZES

Name	US	UK	Φ
boulder	>256 mm	>200 mm	−8
cobble	64–256 mm	60–200 mm	−5 to −6
pebble (gravel)	2–64 mm	2–60 mm	−1 to −6
sand	62.5–2000 mm	60–2000 mm	3 to −1
silt	4–62.5 mm	2–60 mm	8 to 4
clay	<4 mm	<2 mm	8

INTERNATIONAL PARTICLE CLASSIFICATION

Very Coarse Soil

large boulder	>630 mm
boulder	200–630 mm
cobble	63–200 mm

Coarse Soil

GRAVEL

coarse gravel	20–63 mm
medium gravel	6.3–20 mm
fine gravel	2.0–6.3 mm

SAND

coarse sand	0.63–2.0 mm
medium sand	0.2–0.63 mm
fine sand	0.063–0.2 mm

Fine Soil

SILT

coarse silt	0.02–0.063 mm
Medium silt	0.0063–0.02 mm
fine silt	0.002–0.0063 mm

CLAY | ≤0.002 mm |

SOIL PERMEABILITY CLASSES

Class	cm/hour
very slow	less than 0.13
slow	0.13–0.3
moderately slow	0.5–2.0
moderate	2.0–6.3
moderately rapid	6.3–12.7
rapid	12.7–25
very rapid	more than 25

PERMEABILITY OF SOIL TEXTURES

Soil texture	Average Permeability (cm/hour)
sand	5.0
sandy loam	2.5
loam	1.3
clay loam	0.8
silty clay	0.25
clay	0.05

Permeability is a function of particle size and it is particle size that determines soil texture. That means it is possible to determine an average value for the permeability of a soil of a known texture. The second table gives the average permeability for six soil textures.

How to irrigate

Even in rainy Britain, outdoor soils sometimes become too dry and certain plants benefit from irrigation. Greenhouse and other indoor crops depend on the grower for the water they need. In all cases, indoor or outdoor, it is important to supply only as much water as the crops require, and to ensure that it is applied to the root area. There are three reasons for this. The first and most obvious is that it is wasteful to provide more water than is needed, especially during dry weather when public supplies may be under pressure.

The second reason for moderation is that an excess of water will fill too many of the soil pores, thereby expelling air that plant tissues need for respiration. Some tropical plant species that grow in compacted or waterlogged soils have pneumatophores, or breathing roots, to allow them to survive the airless conditions below ground. Pneumatophores are specialized roots that are partly above ground, and the aerial parts have many pores through which the plant exchanges gases. The external pores lead to connected spaces between cells that allow air to diffuse to all parts of the plant. These species are adapted to life in airless soils. Garden plants other than bog and pond species are not, and waterlogged soil will kill them.

A waterlogged soil will dry out, but in very warm weather there is a risk that the very drying will cause harm because it will occur mainly by evaporation,

Sprinkler irrigation supplies water to dry ground prior to planting or sowing. It is effective, but much of the water is lost to evaporation before penetrating the soil.

rather than by drainage. As surface water evaporates, more water is drawn upward through capillary spaces in the soil and evaporates in its turn. It is only the water that vaporizes. Any salts dissolved in the water will remain at the surface, where they will concentrate, possibly reaching levels that are toxic to plants. This is a serious problem in warm climates with irrigation systems that do not include adequate drainage to remove surplus water. The only way to remedy the problem once it happens is to flood the soil to dissolve and wash out the salts.

The condition at the soil surface is a poor guide when deciding whether water is needed. The surface may be dry but the moisture below ground may be sufficient. The place to check is about 30 centimetres below the surface, and some experience with your own soil will help. Clay soil may feel moist, but the moisture may be tightly bound to soil particles and not easily accessible to plant roots, so some watering may be needed, and sandy soil may retain sufficient moisture even though it feels dry. But these are at the extreme ends of soil texture and the feel of most soils is a satisfactory guide.

Sprinkler irrigation systems are used mainly to water lawns and dry ground prior to sowing or planting. They use a great deal of water and apply it at the surface, where much is quickly lost by evaporation or surface runoff. A watering can is more efficient in a small area or a hosepipe in a larger area, and they make it possible to place the water around the base of the stem, from where it will move downward to the root area.

If the garden is too big even for a hosepipe, you may need to install an automatic watering system. Drip irrigation is the most efficient. It consists of pipes or hoses with small holes at intervals. The pipes lie along the ground beside the rows of crop plants and deliver water in small but regular amounts directly to the base of each plant, from where it soaks into the soil around the roots.

Most trees do not need irrigation because their root systems are capable of finding the moisture the plant needs. A basin system is probably the best way of supplying water to those that do, especially in loam soils. A basin is a square or circular area of ground around the tree, usually about 1 metre on each side or five metres diameter, in which the surface slopes downward away from the tree to a shallow trough around the edge. Water poured into the trough saturates the outer part of the basin and moves by capillarity toward the tree and downward at the same time. The water reaches the roots without wetting the trunk. The basin should be enlarged as the tree grows, so its area always matches that of the tree foliage. This is a very economic way to irrigate because it reduces water loss to a minimum. Alternatively, water can be applied to a circular furrow surrounding the tree, without making a basin, and orchard trees can be supplied from furrows, about 45 centimetres wide and 15 centimetres deep, between the rows.

A mulch of organic material will help conserve water. It will increase the amount of moisture sandy soil retains and it will increase the amount of pore space in clay soils. Regular weeding also conserves water, because the weeds transpire moisture just as crop plants do, but unproductively from the grower's point of view.

Sheltering the crop plants from the wind reduces evaporation losses, and the crop itself provides the simplest shelter if the plants are grown in blocks rather than rows.

Plants and climate

It is the climate that sets the conditions in which plants grow and the reaction between the climate and surface rocks that forms the soils, which provide anchorage and supply the mineral nutrients plants need. Over the course of their long history, however, plants have had to adapt to the environments that sustain them. Some of those adaptations have resulted in features that we may consider merely attractive, without recognizing their evolutionary significance.

There are, for instance, temperature limits outside which essential biochemical processes cease. Why is more warmth beneficial but too much warmth fatal? Why do the aerial parts of some plants die down in winter or during a dry season, and why do some woody plants shed their leaves while others retain them? How do plants survive in deserts?

Having described climate, weather, and soils, it is time to consider how plants respond.

Effects of heat and cold

Spring is the time of renewal. Crocus and daffodil leaves push above the surface, the primroses flower, and leaf buds become prominent on the trees. The plants are responding to longer days and a rise in temperature.

Temperature is important because every aspect of plant metabolism comprises biochemical reactions catalyzed by enzymes. Enzymes are molecules of protein or mainly of protein, produced inside cells, which facilitate chemical reactions while remaining unchanged. They do this by colliding with and attaching to substrate molecules, then orienting them in such a way as to facilitate their bonding to other molecules. In order to collide with and affect a substrate molecule, the enzyme molecule must possess more than a minimum kinetic energy, known as the activation energy, and that energy is supplied as heat. As the temperature rises, therefore, enzyme molecules move faster, the likelihood of collisions increases, and once the activation energy is attained, biochemical reactions commence and then accelerate.

Enzymatic reactions are most dynamic between 10°C (50°F) and 30°C (86°F) and within that range their rate increases exponentially by two to three times—200–300%—for each 10°C (18°F) rise in temperature. There are exceptions for reactions in which a physical barrier obstructs access to reaction sites or there are limits to the available substrates. In those cases the reaction rate increases like others with the first rise in temperature but then slows from an exponential to a linear rate.

Photosynthesis is the most important biochemical process for plants, because it converts electromagnetic radiation (solar energy) into chemical energy. The first stage in photosynthesis requires light. It accelerates with increasing light intensity but it is not affected by changing temperature. The second, light-independent stage, on the other hand, comprises a sequence of

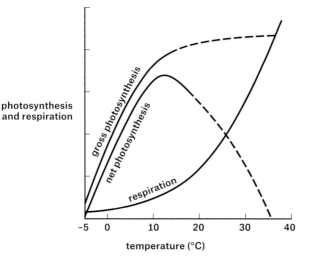

photosynthesis
and respiration

temperature (°C)

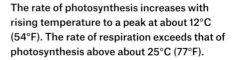

The rate of photosynthesis increases with rising temperature to a peak at about 12°C (54°F). The rate of respiration exceeds that of photosynthesis above about 25°C (77°F).

biochemical reactions and, like all biochemical processes, its rate increases with rising temperature until it reaches an optimum at about 12°C (54°F), after which it declines. In spring, as the sunlight strengthens and the temperature rises in response, photosynthesis accelerates. But the light intensity and temperature continue to rise through the summer, so an additional factor must be contributing to the dramatic burst of spring growth. That factor is the difference between day and night temperatures.

Spring days are pleasantly mild, but as often as not the nights are still cold. The warm days favour rapid photosynthesis and the cool nights do not affect it, because photosynthesis does not take place in darkness. Plants also respire. Chemically, respiration and photosynthesis are mirror opposites. Photosynthesis assembles carbohydrate molecules from carbon dioxide and water, while respiration oxidizes carbohydrate molecules to release energy. Like all aerobic organisms, plants must respire all the time they are metabolically active. During the day, therefore, they utilize in respiration a proportion of the carbon they absorb during photosynthesis, but there is a net gain of carbohydrate because photosynthesis produces more carbohydrates than respiration "burns." At night, however, there is no photosynthetic compensation for respiration, so the plant loses carbon as carbon dioxide. Both processes are affected by temperature, however, and the cold spring nights reduce the rate of respiration. Consequently, a smaller proportion of the net daytime gain in carbohydrates is lost through respiration at night, which leaves more carbohydrates for plant growth than is the case during those balmy nights of summer. The graph on page 213 shows the rate at which photosynthesis and respiration change with rising temperature.

Photosynthesis ceases altogether at temperatures lower than –6°C (21°F) and it is most efficient at temperatures above 5°C (41°F). As spring advances, therefore, photosynthesis accelerates in plants with leaves that are already open and exposed to sunlight. Annual plants survive the winter as seeds, which must first germinate. Many plants, including brassicas, carrots, and other vegetables, will germinate at temperatures higher than 4.5°C (40°F). Others, including tomatoes, beans, and cucumbers, germinate at temperatures above 15°C (59°F). Some require a difference between day and night temperature, usually of about 10°C (18°F) for germination and sturdy growth. Once the summer temperature rises above 25°C (77°F) many spring plants will not germinate at all.

As the temperature rises above 35°C (95°F) the rate of photosynthesis starts to decrease sharply. At these high temperatures proteins denature, which is to say that their structure alters and their biological activity changes or ceases. Photosynthesis ceases at about 45°C (113°F) and few temperate-climate plants can survive temperatures higher than this.

Although there are general temperature limits for photosynthesis, plants vary in the minimum, optimum, and maximum temperatures for growth. The table opposite lists the temperature ranges for a few examples.

How freezing kills

Perennial plants that grow high on mountainsides or in high latitudes must survive long, bitterly cold winters. The plants cannot grow when it is too cold and too dark for photosynthesis, so they become dormant. As the temperature falls, their cells accumulate sucrose and other solutes. This protects the plant tissues by lowering the freezing temperature of the cell solution and is effective until the temperature falls below about –7°C (19°F). To protect against lower temperatures the plant cells produce other compounds which prevent ice forming in and between cells, and that strengthen the cell wall. Woody plants also benefit from the thermal insulation provided by their bark.

Freezing can kill plants in two ways. The first is by dehydration. This is not the dehydration that results when the ground contains too little moisture, which can occur at any time of year, but dehydration of the plant tissues themselves, due to osmosis.

Osmosis is the process by which solutions of different strengths equalize

TEMPERATURE RANGE FOR SELECTED PLANTS

Plant group	Low-temperature limit (°C)	Optimum temperature (°C)	High temperature limit (°C)
agricultural crops	−2 to 0	20 to 30	40 to 50
desert plants	−5 to 5	20 to 35	45 to 50
temperate annuals, spring bulbs	−5 to −2	10 to 20	30 to 40
tropical trees	0 to 5	25 to 30	45 to 50
temperate deciduous trees	−3 to −1	20 to 25	40 to 45
evergreen conifers	−5 to −3	10 to 25	35 to 42
heath and tundra shrubs	about −3	15 to 25	40 to 45
temperate mosses	about −5	10 to 20	30 to 40
lichens	−10 to −15	8 to 20	25 to 35

when they are separated by a semipermeable membrane, such as the cell wall of a plant. Molecules of a solvent are smaller than those of the solute dissolved in it and the semipermeable membrane allows small molecules to pass, but blocks large ones. In plants the solvent is water and there is a solution both inside and outside the cell. If the solutions inside and outside the cell are at the same strength, an equal number of water molecules will pass through the cell wall in either direction, so although water is constantly moving, the concentration inside and outside the cell remains constant. The cell is then said to be isotonic. If the solution inside the cell is more concentrated than that outside, the cell is hypotonic (hyposmotic) and water molecules will flow into the cell, making it swell and become rigid, or turgid. If the solution outside the cell is the more concentrated, the cell is hypertonic (hyperosmotic) and water will flow out of it. This process is called plasmolysis and it causes the cell to shrink and become flaccid. The diagram on page 216 shows these three states.

When an aqueous solution freezes it is the solvent, water, which solidifies, removing solvent but leaving solute behind. The presence of a solute alters the freezing temperature. Depending on the concentration, a solution of common salt (NaCl), for example, lowers the freezing temperature to about −16°C (3°F) and seawater, with an average salt concentration of about 3.5%, freezes at −1.91°C (28.56°F). As a solution freezes, its volume decreases through the solidification of solvent, and its concentration increases.

The solution in and between plant cells lowers the freezing temperature,

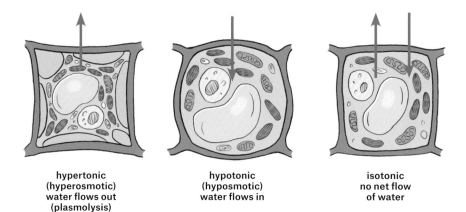

| hypertonic (hyperosmotic) water flows out (plasmolysis) | hypotonic (hyposmotic) water flows in | isotonic no net flow of water |

If a plant cell is hypertonic, the solution outside the cell is stronger than that inside; water flows out of the cell, which shrinks (plasmolysis). If the cell is hypotonic, the solution inside the cell is stronger than that outside; water flows into the cell making it turgid. If the cell is isotonic, the solutions inside and outside are equally strong; water flows in both directions.

but by only 1–2°C (2–4°F) and in unhardened plants the tissues start to freeze when the temperature falls below –1°C (30°F) in the presence of freezing nuclei. These are always available in the form of bacteria and fragments of cell debris. Hardening prepares the plant for low temperatures by lowering the temperature at which freezing occurs or helping the plant tolerate cold.

If the freezing occurs mainly in the extracellular solution, as the concentration of that solution increases water is drawn out of the cells by osmosis. The cells become hypertonic and dehydrated and non-woody tissues become flaccid. Water that leaves the cells also freezes, so after a time the aggregations of ice crystals outside the cells grow much larger than the cells themselves. This continues until the cells are severely desiccated, collapse completely, and die.

If freezing occurs inside the cells, the cells become hypotonic and water flows into them, only to freeze onto the ice that has already formed. This continues until the cells burst.

Ice crystals also damage the cell wall and contents directly while they are forming. The extent of the damage freezing causes depends to some extent on the speed with which it happens. It is more likely that the cell contents will freeze if freezing is rapid. If it occurs more slowly, the freezing is more likely to be intercellular and the cells have a better chance of rehydrating and recovering when the ice thaws.

HARDINESS RATING

Rating UK	Rating US	Temperature range (°C)	Category	Details
H1a	13	>15	tropical glasshouse	under glass all year
H1b	12	10 to 15	subtropical glasshouse	outdoors in favoured locations in summer
H1c	11	5 to 10	warm-temperate glasshouse	outdoors in summer
H2	10b	1 to 5	tender, unheated glasshouse	cannot survive being frozen
H3	9b/10a	1 to –5	half-hardy, unheated glasshouse	hardy in places with mild climate or microclimate
H4	8b/9a	–10 to –5	hardy, tolerates most winters	may suffer in cold places or wet soil
H5	7b/8a	–15 to –10	hardy, tolerates cold winters	may suffer in exposed sites
H6	6b/7a	–20 to –15	hardy, tolerates very cold winters	hardy in most places, but container plants need protection
H7	6a/1	<–20	very hardy	hardy everywhere

Obviously, some periods of low temperature are more serious than others. In a mild frost the temperature in temperate climates falls to about –2°C (28°F) and remains there for no more than about two hours. Ice may form on the exterior of plants, but only the most tender plants are harmed. A hard frost lasts for several hours with temperatures of –4°C to –2°C. It damages plant tissues in leaves and blossoms, mainly by making them hypotonic until they burst. If the temperature remains below –4°C (24°F) for several hours, it is a severe freeze and cells will suffer desiccation.

Hardiness

Many tropical and subtropical plants will be injured or even killed at temperatures that are cool but well above freezing. Plants native to temperate regions may tolerate hard winters, but yet suffer in late frosts that catch them when their leaf or flower buds are opening. The ability of a plant to withstand cold is known as its hardiness, and it varies widely, reflecting the climates in which species have evolved.

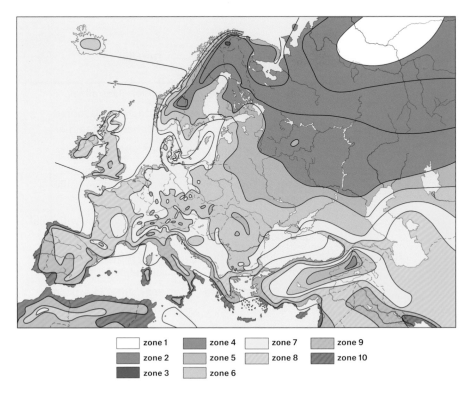

	zone 1		zone 4		zone 7		zone 9
	zone 2		zone 5		zone 8		zone 10
	zone 3		zone 6				

Hardiness zones in Europe, labelled according to the US classification.

Plants that are adapted to cold winters have ways of avoiding the dangers. High-latitude conifer forests are often very dense. The trees shade the ground, making the forests dark, mysterious, confusing places of legend and folk tales. But the trees are also partly shading themselves, and shading reduces the rate at which they lose heat by radiating it upward. The branches that shade those below also tend to hold and retain snow, again reducing heat loss.

Large size is also a way to conserve warmth. A large plant or structure such as a bud or fruit has a small surface area in relation to its volume. The bulk of the structure absorbs warmth but the small surface area restricts the rate at which the warmth is lost by radiation.

Plants also have internal features that help them tolerate cold. Dehydration leading to desiccation is possibly the most harmful consequence of freezing. It is due to the flow of water out of cells by osmosis, and a way to prevent or at least reduce it is to alter the osmotic pressure inside cells. Some plants achieve this by increasing the concentration of solutes in the cell protoplasm.

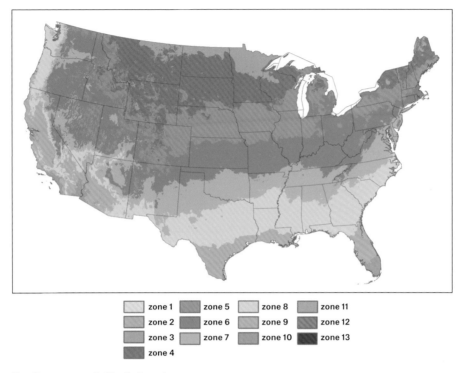

Hardiness zones in North America.

The extracellular solution freezes, raising its concentration, but the high solute content on the other side of the cell membrane prevents the cell becoming hypertonic and the ice crystals outside from growing at the expense of cellular fluid. The plant tolerates freezing by conserving cell moisture. This technique allows halophytes (salt-tolerant plants) to thrive in brackish water and many halophytes are very hardy. Other plants, such as many willows (*Salix* species), paper birch (*Betula papyrifera*), and trembling aspen (*Populus tremuloides*), are able to rehydrate their collapsed cells when the extracellular ice thaws. In many plants, fluids can be supercooled (chilled to below freezing temperature while remaining liquid). This avoids the dangers of freezing altogether.

As the cold season approaches, marked by the changing day length, plants harden themselves. They may accumulate sugars or sugar alcohols in their tissues, lowering the freezing temperature inside the plant. They may increase the supercooling of tissue fluids, or produce lipids to toughen their cell walls to help them survive desiccation.

Although many plants are able to harden in preparation for winter, a brief spell of mild weather can undo their hardiness. Once winter has taken hold and the plants have shut down, a rise in temperature can be harmful, pleasant though it may be for us. Some growers protect their plants at such times by spraying them with water during the day just to keep them cool.

Water can also help plants through a cold spell. Heavy watering the day before a hard frost is forecast allows the plant to take up enough water to dilute its tissue solutions significantly. Once the temperature has fallen below freezing, misting the plant will insulate it. The mist will freeze, covering the plant with a thin coating of ice. The ice will be too thin to damage the plant, but it will greatly reduce further heat loss.

Plant hardiness varies widely, but it has been classified. The most widely used classification was devised by the United States Department of Agriculture, and the Royal Horticultural Society has produced a classification appropriate for plants growing in the British Isles. The table on page 217 displays both classifications, with the temperatures they describe and the way plants in each category should be protected.

The hardiness ratings can also be interpreted geographically, identifying areas in which the designated temperatures can be expected. This leads to the concept of hardiness zones, and the two maps (on pages 218 and 219) show these zones, the first in Europe and the second in the United States.

Soil temperature and germination

Plant seeds vary greatly in size and appearance, but they all contain some or all of the same principal ingredients, shown in the illustration (opposite) of a typical seed of a dicotyledon (dicot) plant. There is an outer protective testa, or seed coat, enclosing nutrients to sustain the seedling, and the embryo of a new plant consisting of one (monocots), two (dicots), or two or more (gymnosperms) cotyledons or seed leaves, a hypocotyl, which is part of the shoot, lying above the radicle, or embryonic root.

Seeds form toward the end of the growing season and remain dormant through the winter, germinating when winter has ended and the environmental conditions are favourable to the young plants. If you cut open a wheat seed you will find it is very dry. Other plants have seeds moister than this, but all seeds are fairly dry and they need to absorb water before the embryo can begin to develop. The process is called imbibition. Seeds swell as they absorb water, bursting the testa and bringing the contents of the seed into contact with the soil and the soil solution. At the same time, moistening the nutrient store inside the seed activates enzymes, which catalyze the biochemical reactions

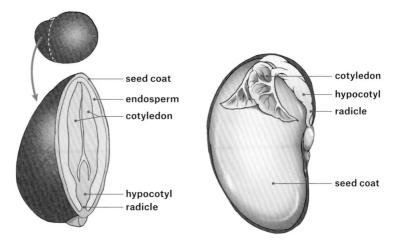

The structure of a dicot seed showing: the inner and outer layers of the testa (seed coat); endosperm; cotyledon (of which there are two); and hypocotyl. The radicle is below the hypocotyl.

that start converting nutrients into compounds that are metabolically useful to the young plants.

Moisture is essential for germination, therefore, but so is oxygen, because until the seed becomes a seedling, with leaves above ground, the young plant is unable to photosynthesize, relying instead on aerobic respiration to release energy from its carbohydrate store. Air fills most of the pore space between soil particles, so the seed is able to exchange gases either through its testa or, if this is impermeable, directly with the seed contents once the testa has broken open. If the soil is saturated, however, water will fill most or all of the pore space, driving out the air. Respiration then being impossible, the seed will remain dormant and it will die if the soil remains saturated. So water is necessary, but in moderate amounts.

Enzymes cannot function until they have the requisite activation energy, and that means they remain inactive while the seed lies in the cold soil of winter and they must be warmed before the new plant can develop. The temperature at which seeds germinate varies widely, but for all of them there is a minimum temperature, below which they will not germinate, and an optimum temperature for germination. Most will not germinate at temperatures higher than the upper end of the optimum range. The table on page 222 lists these temperatures for a number of vegetables.

Some seeds have additional requirements. Many plants of temperate climates will not germinate unless they experience the prolonged low temperatures of winter. Exposure to cold ensures that they produce fruit and seeds in

MINIMUM AND OPTIMUM TEMPERATURES FOR GERMINATION

Crop	Minimum (°C)	Optimum (°C)
aubergine	16	24 to 32
bean	8 to 10	16 to 30
beet	4	10 to 30
cabbage	4	7 to 35
carrot	4	7 to 30
cauliflower	4	7 to 30
celery	4	15 to 21
cucumber	16	16 to 35
lettuce	2	4 to 27
maize (corn)	10	16 to 32
onion	2	10 to 35
parsley	4	10 to 30
parsnip	2	10 to 21
pea	4	4 to 24
pepper (capsicum)	16	18 to 35
pumpkin	16	21 to 32
radish	4	7 to 32
rutabaga	4	16 to 30
spinach	2	7 to 24
squash	16	21 to 35
tomato	10	16 to 30

spring or summer, allowing the seeds to germinate before the onset of winter. Winter-sown cereals fall into this category and growers often chill the seeds at 5–10°C (40–50°F) prior to sowing. The process is called vernalization and it encourages the seeds to germinate in late autumn, remain dormant through the winter, and resume growth in early spring, giving them an advantage over spring-sown varieties. Fruit trees need cold weather to trigger dormancy at the start of winter and a second exposure to cold to bring them out of dormancy.

Wet soil feels cold, even on a warm day. That is because it really is cold. Water has a much higher heat capacity than the mineral grains that make up most of any soil. It takes 4.19 joules of energy to raise the temperature of water by 1°C (1.8°F), compared with the 0.8–0.9 joules needed to warm mineral grains

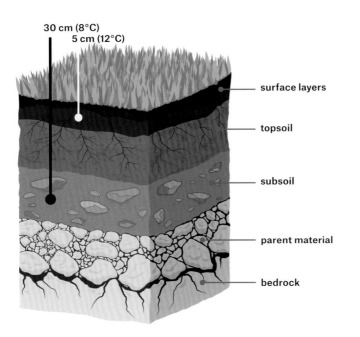

30 cm (8°C)
5 cm (12°C)

surface layers

topsoil

subsoil

parent material

bedrock

Soil is a good thermal insulator and while the temperature near the surface fluctuates, it remains much more constant in lower horizons, cooler in summer and warmer in winter. This shows the contrast in temperatures 5 centimetres and 30 centimetres below the surface of soil on a summer day.

by the same amount. It is important, therefore, to check the soil temperature where you plan to sow seeds, because it may not be the same everywhere in the garden. Moist clay soil will be cooler than dry, sandy soil.

Soil is a good thermal insulator. Scientists compiling temperature records sometimes take measurements from boreholes in soil because changes penetrate only very slowly, so a soil profile may provide an annual temperature record going back over many decades or sometimes centuries. As the diagram above illustrates, the temperature in the subsoil changes very slowly, so on a summer day, 30 centimetres from the surface it may be at 8°C (46°F) when the temperature close to the surface is 12°C (54°F).

How the wind stunts plant growth

Windstorms often tear off branches and bring down whole trees, especially if they strike deciduous trees that are in full leaf, because the leaves offer wind resistance, like sails. So, obviously, wind can and does damage plants. But the effects of gentler winds are more complex.

As it sways in the wind a woody plant releases a growth hormone that stimulates its production of lignin, the polymer that cements cellulose fibres in cell walls, thereby strengthening them. As a result, plants that are repeatedly exposed to wind have shorter, thicker stems than plants growing in sheltered locations. The effect of physical contact on the overall shape and appearance

This tree appears to be bending in the wind, but the upright grass shows the air is still. The wind has sculpted the tree to this shape by drying and killing the shoots and leaves exposed to it, while the tree itself shelters the branches on the downwind side.

of a plant—its morphology—is called thigmomorphogenesis and herb growers exploit it by brushing their plants almost flat every day for the first two to three weeks after planting. This is similar to the effect of the wind and it produces plants with stronger, stockier stems, thus improving their ability to support the weight of the plant.

Trees and shrubs growing in exposed sites often have a windswept look. They appear to lean downwind as though the wind is bending them over. Near the coast, the plants usually bend away from the sea. That they are not actually being blown to the side becomes evident when the air is still, because they retain their windswept shape even when the wind isn't blowing. In fact, the

PLANTS AND CLIMATE

trees and shrubs are not bending with the wind. They have been sculpted by it.

Plants release and absorb gases through their leaf stomata and for this to happen the stomata (pores) must be open. While the stomata are open, however, moisture in the plant tissues is able to evaporate through them. This is transpiration and the loss of water draws more water upward through the xylem, so there is a constant upward flow. Immediately adjacent to the leaf surfaces there is a boundary layer of air that is close to saturation because of the moisture being transpired into it, and the high moisture content in the boundary layer constrains the rate of transpiration.

A strong wind can sweep away that boundary layer, however, replacing it with much drier air and thereby removing the constraint on transpiration. The rate of transpiration accelerates and it may increase to such an extent that the plant loses moisture faster than rising moisture can replace it. The effect is exacerbated near the coast because onshore winds carry salt crystals that readily absorb water, intensifying the drying effect. Plant cells cannot survive without water and tissues exposed to a strong prevailing wind wilt and die back.

Desiccation affects only the upwind side of the plant, because the plant itself shelters the tissues on the lee side. Consequently, the tree or shrub dies back on one side but grows normally on the opposite side, with the result that it appears to lean downwind.

Wind sculpting is most dramatic in marginal environments, close to the limits for tree growth. There, in high latitudes and high on mountainsides, woody plants are often coated in snow or ice. This protects them, but a dry wind bringing air that is above freezing temperature will melt the ice, exposing the plant to the drying wind, at a time when the roots are in soil which is at or close to freezing and are able to absorb only limited amounts of water. These conditions produce stunted, gnarled shrubs and trees, some of which grow close to the ground. Such vegetation is called krummholz.

Why deciduous plants shed all their leaves

In most climates there is a time of year when plant growth ceases. This may either be the cold time, the winter of high latitudes, or the dry season of lower latitudes—summer in some places, winter in others. Even at the equator where the day length barely changes through the year, many areas experience a dry season. Cold or dry, the challenge is the same. There is insufficient water. In cold climates this is because the water around plant roots is frozen so the roots are denied access to it. Perennial plants have no option but to shut down.

While they remain dormant, plants must maintain their own tissues and

Krummholz, also called elfin wood, comprises gnarled, stunted trees growing high on a mountainside between the forest line and the tree line.

The brilliant fall colours of a deciduous forest develop as the trees cease to produce chlorophyll and some produce red pigments to confuse aphids.

leaves are a problem. These are vital organs, but they are exposed on the exterior of the plant, making them vulnerable to winds and in high latitudes to freezing temperatures. Plants have evolved several strategies for coping with this difficulty and one of them brings us the spectacular colours of the New England fall and British autumn. They produce disposable leaves.

To our modern ears, attuned to the virtues of thrift and recycling, throwaway leaves sound wasteful, but the strategy works well in climates where winter temperatures fall to about –30°C (–22°F) and rise to about 30°C (86°F) in summer, giving a growing season that lasts five or six months. And it is less wasteful than it seems.

Since the leaves need to last for no longer than six months they need not be robust, but they do need to grow rapidly in spring. So the leaves are thin and flimsy, but with a large area. In terms of energy and resources they are cheap to produce.

As the autumn days grow shorter and the temperature falls, the plants begin to prepare for winter. They reduce production of auxins, the hormones that promote cell growth, and increase production of ethylene, which inhibits growth. They cease production of chlorophyll and nutrients begin to flow out of the leaves and back into the body of the plant where they will be stored until spring. As the chlorophyll breaks down and is not replaced, the leaves lose their green colour. Most turn yellow, but that is not really a good colour for a leaf, because aphids find it attractive. So some plants produce red pigments and direct them into the leaves to confuse the aphids. Eventually the yellow leaves lose that coloration and turn brown. Hence the succession of green, yellow, brown, and red colours that delight us, and if the weather in a

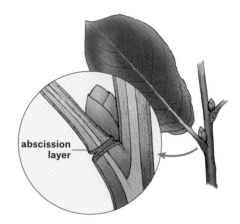

abscission layer

At the base of every leaf petiole there is a layer of tissue, the abscission zone, that weakens when it is time to shed the leaf until the slightest gust of wind will break it, allowing the leaf to fall.

particular summer compresses the autumn season, as sometimes happens, the plant processes accelerate and all the colours are visible at the same time.

What remains is leaf tissue that has lost its nutrients and chlorophyll. Its cells are dead and it is of no further use, so the plant discards it. The process of discarding is called abscission. Most leaves grow on a short stalk, called a petiole, which attaches it to the plant stem. Near the point where the petiole—or the leaf base if there is no petiole—joins the stem there is an abscission layer, sometimes visible as a slight swelling. This is a double layer of parenchyma tissue, comprising unspecialized cells with air spaces between them. On the stem side of this layer, on the upper side of the point where the stem joins the main branch or stem, there is a small bud that will produce next year's leaf. The above illustration shows this arrangement.

Cells and the spaces between them in the part of the abscission layer closest to the stem fill with fatty material, forming a seal. Cells in the outer layer are short, with thin walls, and the layer weakens when its nutrients are withdrawn. Eventually it can no longer support the weight of the leaf and a slight breeze or shower of rain is enough to complete the process. Leaves that are shed all at the same time, as well as the plants that shed them, are said to be deciduous.

Even then, what remains of the leaves is not lost. The dead leaves fall to the ground, where they decompose fairly rapidly, feeding a hierarchy of organisms and returning nutrients to the soil.

Why evergreen plants retain their leaves

Deciduous leaves work well where there is enough time during the growing season to unfurl the leaves and still photosynthesize for long enough for the plant to grow and produce seed. A deciduous tree needs 120 days when the temperature is above 10°C (50°F), and precipitation distributed evenly through the year.

Not all plants enjoy such an advantage and these have evolved an alternative strategy. Evergreen coniferous trees thrive with a growing season of no more than 30 days and they do it by using every glimmer of sunlight they can. That is why they are evergreen.

An evergreen plant is one that retains leaves throughout the year. That is not to say it does not shed it leaves. Leaves wear out, become damaged, work less efficiently, and the plant must replace them with fresh leaves. But it does not replace all of them at the same time. With the exception of larches (*Larix* species), *Taxodium* species (bald cypresses), *Metasequoia glyptostroboides* (dawn redwood), and *Ginkgo biloba* (ginkgo), all of which are deciduous, coniferous plants shed leaves throughout the year, but just a few at a time, so they are always in more or less full leaf.

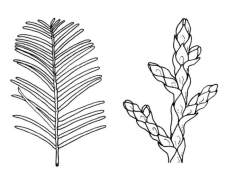

Two types of conifer leaves, of pine (left) and western red cedar (right).

Since the leaves must last much longer than deciduous leaves they need to be more robust. They are tough and they are also small. They remain active through the dry season, so they must conserve moisture. Their small size helps with this, but so does their waxy cuticle, or outer skin. The leaves have fewer stomata than those of deciduous leaves and the stomata are located in crevices or pits that shelter them from the wind, reducing water loss by transpiration. The leaves themselves are in the form either of long, narrow needles or tiny, overlapping scales. *Pinus* species (pines) produce long needles and *Cupressus* species (cypresses) and *Thuja plicata* (western red cedar) have scale leaves. These are shown here in the illustration.

With leaves that are present and contain chlorophyll all year, conifers are able to photosynthesize whenever the sunlight is sufficiently intense, even on mild days in the depths of winter. In spring, the plants do not have to wait while they grow leaf buds and unfurl their new leaves, but are ready to begin photosynthesis the moment there is enough light and warmth. Long winters and cool summers do not inconvenience them.

Conifers also have a reproductive advantage over many broad-leaved species. Although it takes them much longer to produce seeds—it takes several years for a female cone to mature—rather than soft fruit, the seeds are held in tough cones that can withstand cold and wind, rain, snow, and ice.

You can grow coniferous trees and shrubs almost anywhere. They are adapted to harsh winters and short summers but thrive in milder conditions. That is less true of broad-leaved deciduous species, which are more restricted. And the difference between their seasonal strategies explains why broad-leaved, deciduous trees form forests in warm

Coniferous forest in winter.

continental and humid subtropical climates, while the great belt of coniferous forest, the boreal forest or taiga, occupies the subarctic and cold continental regions of northern North America and Eurasia.

Why spring flowers flower in spring

Many of our most popular flowers appear in early spring. Primroses, crocuses, daffodils, snowdrops—the list is quite a long one and it includes a few woody plants such as blackthorn and cherry plum. We cultivate the herbs, of course, but many also grow wild, along roadside verges and, most of all, in woodland.

All living organisms need to reproduce and that primary requirement determines most aspects of the way they live. Flowering early in the year gives certain plants their best chance of producing viable seed, but the strategy works only in broad-leaved, deciduous woodland.

Before they can set seed, plants must emerge from winter dormancy, grow, produce flowers, and attract pollinators, and they can do none of these things without adequate light for photosynthesis. In early spring the days are lengthening and the sunlight is becoming steadily more intense. Leaf buds are swelling on the deciduous trees, but it will be a little while before the leaves open. During this interval there is no leaf canopy to shade the ground and the strengthening sunlight falls directly on to it. There is abundant light.

Plants also need moist soil. During the winter, snow may have covered the forest floor. As the temperature rises the snow melts, moistening the soil. If there was no snow, winter rainfall and spring showers provide moisture, and at a time when although temperatures are rising they are not yet high enough

Spring flowers flourish briefly on the floor of a deciduous forest, between the time the temperature rises sufficiently for photosynthesis and the tree canopy closes, shading the ground.

for the rate of evaporation to be high enough to dry out the soil during spells of dry weather. There is abundant moisture.

The plants are able to grow and to produce flowers. Now the flowers must be pollinated. Some flowers are wind-pollinated. While the trees remain bare, the wind encounters less resistance. It blows more freely through the forest and is able to transport pollen. This is a benefit to birches (*Betula* species), alders (*Alnus* species), hazel (*Corylus avellana*), and other trees that produce flowers in catkins. Once the leaves emerge they will trap much of the wind-blown pollen, preventing it from reaching the sticky pistils of the female flowers.

Insects pollinate most woodland flowers, however, and the absence of leaves also benefits them. The flying insects have longer lines of sight and fewer obstructions to negotiate. Here, though, the plants must compete for pollinators, for there are many of them and the time available for pollination is brief. So they produce showy yellow, pink, and blue flowers that scream for attention, albeit silently. Then, once the flowers are pollinated, the plants produce seeds, send nutrients to storage organs such as bulbs and corms, and before long, their work done, they die down as the trees come into leaf and spring moves into summer. There are also woodland plants that flower in summer, but most of those produce white flowers, which are more visible on the dimly lit forest floor.

Blackthorn (*Prunus spinosa*), with wood once used to make walking sticks and shillelaghs and fruits (sloes) used to flavour gin, flowers early, but cherry plum (*P. cerasifera*) is still earlier, often flowering in late February. These are insect-pollinated woodland shrubs and small trees that take advantage of the open canopy. Hawthorn (*Crataegus monogyna*) flowers later, in May, at the woodland edge and planted in hedgerows, and marks the beginning of summer.

Why succulents thrive in deserts

The Mojave Desert in California and Arizona—the map opposite shows its location in the rain shadow of the Sierra Nevada—receives less than 125 mm (5 in.) of precipitation in an average year, some of it as snow. It is a rocky, dusty, and very dry place that includes Death Valley, where the day temperature has been known to rise to 57°C (135°F). Death Valley was named in 1849, the year a party of 30 migrants used it as a shortcut to the California gold fields and 12 of them perished. The desert is spectacularly beautiful but harsh, yet many plants thrive there, about 200 of them endemic. The most famous and abundant is the Joshua tree (*Yucca brevifolia*), a yucca that grows more than 10 metres tall at up to 10 centimetres a year.

The young Joshua tree has tender leaves that are attractive to herbivores, but the mature leaves are tough, sharply pointed, and indigestible. They are also concave and when it rains they channel water to the stem, where it runs down to the ground and thence to the roots.

That is one way to make the best use of a restricted water supply but there are others. Almost all cacti are succulent plants. That is to say, they store water in their tissues and it is this that gives them their characteristically swollen appearance.

Cacti store water in large cells in their stems. When cells engaged in photosynthesis are short of water it passes to them by osmosis from the storage cells, which replenish their supply from the shallow roots the next time it rains. Using water causes the storage cells to shrink and they swell again when replenished. Many cacti have strongly ribbed stems, which allow them to expand and contract more easily. If the stem appears smooth and almost without ridges,

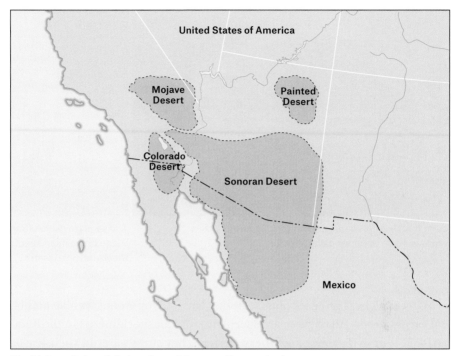

The Mojave, Painted, Colorado, and Sonoran Deserts in the western United States and northern Mexico lie in the rain shadow of the North American Cordillera.

it means the plant is full of water. In most species the leaves are so reduced as to be barely visible, but cacti are renowned for their spines, which are modified leaves borne singly or in clumps. Not all cacti possess spines and some have them only when young, but where they do occur they emerge from sunken cushions called areoles, which are condensed lateral branches or shoots. Flowers also emerge from areoles.

Apart from 35 species of *Rhipsalis* (mistletoe cacti) found in Africa, Madagascar, the Seychelles, and Sri Lanka, all cacti are native to the Americas, although they have become naturalized in other parts of the world. But some of the African spurges (Euphorbiaceae) resemble them closely. These have upright, succulent stems, which in many species are ridged and bear spines. The spines always occur in pairs and they do not arise from areoles, however, so it is not difficult to distinguish euphorbias from cacti.

Spines deter animals attracted to the stored water, but both cacti and euphorbias have a second line of defence. The liquid they store—clear in cacti and a milky latex in euphorbias—is poisonous.

The spines on this *Opuntia* cactus are modified leaves that grow from sunken cushions called areoles.

The saguaro cactus (*Carnegiea gigantea*) is one of the most spectacular plants of the American deserts.

Many gardeners grow ice plants, stone plants, living stones, pebble plants, and carpet weeds. All of these are succulents and most originated in the Kalahari and Namib Deserts of southern Africa. They store water in their leaves and living stones or pebble plants (*Lithops* species) resemble the stones among which they grow so closely they are difficult to see except when they produce their showy flowers. The leaves of ice plants are covered with small, white papillae that give them a frosted appearance.

What happens when the desert blooms

Succulents store water to see them through the long intervals between desert showers, but annual plants have an alternative strategy that allows them to thrive in all but the driest of deserts. They deal with drought by avoiding it altogether.

Annual plants see out the hard times as seeds, dormant and lying securely in the soil. In temperate climates seeds need wait only until rain and melting snow moisten the soil and spring sunshine warms it, but in the desert they have to do rather better than that, because sometimes it doesn't rain for more than a year. Seeds of the Californian desert evening primrose (*Oenothera californica*) remain viable for more than 50 years and many desert annuals have seeds that remain viable for more than 10 years. They achieve long-term viability by slowing their metabolism almost to a standstill and by having tough coats that are watertight—not to prevent water entering, but to prevent its loss.

When rain does fall it soaks into the ground but then evaporates from the surface and is soon gone. There is a real danger a seed that germinates when

PLANTS AND CLIMATE

The flapjack plant (*Kalanchoe thyrsiflora*), with other names including paddle plant, desert cabbage, and white lady, is a succulent native to the deserts of southern Africa but widely distributed in Africa, Madagascar, and Asia, and widely cultivated. It is a **CAM** plant.

moistened may grow into a plant that dies before flowering for want of water. So the seed must germinate promptly and grow fast.

But what if the shower is a false alarm, a brief sprinkling of water that is gone before the plant can set seed? How can a seed know what to expect?

There are ways. Many seeds have chemical compounds that toughen the seed coat, thereby inhibiting germination. These inhibitors are soluble in water, so if the seed remains wet enough for long enough the inhibitors dissolve away and the seed germinates. Seeds will germinate only if the amount of rain is sufficient not simply to moisten the soil but to soak the seed thoroughly.

WHAT HAPPENS WHEN THE DESERT BLOOMS

Many plants produce seeds with coats containing substances on which bacteria or fungi can feed, weakening the seed coat as they do so. The bacteria or fungi require moisture, so in desert soils they will not help germination unless the soil is moist.

Other seeds hedge their bets. *Cassia obtusifolia* (wild senna or sicklepod), a Saharan plant used in herbal medicine, employs this strategy. It produces two types of seeds. One type germinates as soon as the coat is moistened. This gives them a quick start in life and if the rain continues they have an advantage over the seedlings of more cautious competitors. If the rain that moistened them was no more then a brief shower, then the seedlings will die, but the plant does not suffer because it has a second type of seed that germinates only if it is soaked, which means the soil must remain wet for much longer.

Neurada procumbens, a prostrate herb found from the Arabian Peninsula to the Indian Desert, bears seeds that germinate one at a time. Each shower of rain sees the germination of just one seed. That way there is a good chance that at least one seedling will survive.

Successful germination is only the first hurdle. Desert annuals must then grow rapidly in order to set seed before the necessary moisture disappears. *Boerhavia repens* (spreading hogweed) of southern Africa and the Near East may be the fastest. It can grow, flower, and set seeds within eight to ten days of germinating. *Fredolia aretioides* (pillow cushion plant) of North Africa commences photosynthesis within ten hours of germinating, and annual species of *Convolvulus*, native to the northern Sahara, complete their life cycle in six weeks or less.

Where rainfall is more reliable, though sparse, yet another strategy is open to the annual plants. The Mojave Desert has two rainy seasons, for instance, in winter and summer, and there are two groups of annual plants, one that grows in winter and one that grows in summer. Provided there has been at least 15 mm (0.6 in.) of rain, germination is triggered by cool and warm temperatures, respectively.

Germinate promptly and grow fast, but then another problem emerges. Most of the desert annuals rely on insects for pollination, and with so many plants flowering at the same time there is intense competition for pollinators. To address this, the plants have evolved big, bright, showy flowers to catch the eyes of flying insects with a promise of nutritious rewards. And that is why a rare soaking is followed within days by the transformation of the desert into a vast carpet of brilliant flowers.

Desert annuals are very attractive, but if you're thinking of growing them

Within days of a heavy shower, the desert is carpeted by brilliant flowers, produced by plants that must germinate, flower, and set seed before the ground dries out.

you should be aware of their tendencies. To survive in the desert they are tough, resilient, and more than equal to most of the obstacles life can place in the way of their growth and reproduction. Or, to put that another way, they can be highly invasive.

Why some seeds remain viable for decades while others don't

In 2005 scientists induced a seed of the Judean date palm (*Phoenix dactylifera*) to germinate. Archaeologists had recovered the seed from a number preserved in a jar in the palace of Herod the Great at Masada, Israel, and it was about 2000 years old. By 2010 the resulting sapling was about 2 metres tall. In 2007, seeds of narrow-leaved campion (*Silene stenophylla*), radiocarbon dated at 31,800 ± 300 years old, germinated after being recovered from squirrel burrows 38 metres below Siberian permafrost.

These are highly exceptional. Some seeds can be stored under controlled conditions for decades or even centuries, but there are many that cannot be stored for more than a year, or in some cases for more than a few months. A seed that remains viable after storage for a long period is described as orthodox. One that cannot survive desiccation—so cannot be stored—is said to be recalcitrant. Many tropical plants produce recalcitrant seeds—but so does *Aesculus hippocastanum*, the horse chestnut. Such plants are conserved by growing them.

Orthodox seeds also withstand freezing. Seeds that are to be stored in a seed bank are first dried until their water content amounts to 4–5% of their

The entrance to the Svalbard Global Seed Vault, excavated in the permafrost of the mountains of Svalbard, Norway, where duplicates of seeds from collections all over the world are stored to preserve biodiversity and as a resource that will allow those collections to be restored should they be lost through natural disaster, war, or any other cause.

mass, then cooled to 0°C (32°F) if they are to be germinated within the next year or two, or to –18°C (0°F) if they are to be stored for decades. The seeds are then held at that temperature. Seeds readily absorb moisture from the air, so they must be stored in a dry place. About 75–80% of flowering plants produce orthodox seeds.

It is, of course, quite simple to collect seeds from your own plants, dry them, and store them in a dry place for sowing next season. Seed banks are used to conserve endangered species and heritage varieties.

There are many shades of recalcitrance, the less extremely recalcitrant seeds being classed as intermediate. All seeds are dehydrated to some extent before being shed. Depending on species, the water content of a mature seed is usually between 30% and 80% and some tolerate further drying better than others. There are also wide variations in tolerance of chilling.

Seeds that tolerate dehydration also withstand mechanical stress, and to increase their resistance many reduce the size of their cell vacuoles, which ordinarily are filled with fluid, or fill the vacuoles with insoluble substances. The cytoskeleton, which maintains the structure of the cell, becomes dismantled during dehydration. In orthodox seeds the cytoskeleton reassembles during imbibition, but in recalcitrant seeds it fails to do so. DNA also suffers damage during dehydration but in orthodox seeds the damage is repaired on imbibition.

A seed contains a new plant and must ensure its safe survival until the conditions are favourable for its growth. The seed appears to be dry and inert, but it is not dead. Respiration continues, but at a greatly reduced rate and the plant's metabolism is effectively shut down, but ready to revive. Shutting down the metabolism involves a complicated sequence of cellular processes that accompany dehydration. In recalcitrant seeds these cause fatal damage.

Most plants that produce recalcitrant seeds live in warm or wet habitats where the growing season is long and conditions are right for germination as soon as seeds are shed. There is no need for them to wait. Orthodox seeds dehydrate as they mature, but since recalcitrant seeds are to germinate rapidly there is no need for dehydration. The seeds are not fully matured and have a moisture content that is highly variable, but usually greater than 30%, which is the minimum level needed for germination. The plant sheds them at intervals and they germinate almost at once. The seeds are usually large, with the cotyledons accounting for almost all of the mass, protected by a thin seed coat and a hard pericarp, and most recalcitrant seeds occur in single-seeded fruits.

Why many Mediterranean plants have tough, leathery leaves

Not all broad-leaved plants are deciduous. Broad-leaved evergreens, such as holly (*Ilex* species), holm oak (*Quercus ilex*), and many rhododendrons, are widespread and fairly common, and most trees in tropical rain forests are also broad-leaved evergreens. Deciduous plants shed their leaves during a dry or cold season, when water and nutrients are less readily available. Broad-leaved evergreens are less tolerant of extreme cold, so most broad-leaved plants of temperate latitudes are deciduous, but in warmer climates retaining leaves is an adaptation to low nutrient levels, in some cases due to acid soils. Evergreens thrive on poor soils and their fallen leaves contain more carbon in relation to nitrogen than those of deciduous plants, a ratio that increases soil acidity, so once they become established, evergreens may thrive by creating conditions less favourable to deciduous species.

Evergreen leaves do fall, of course, but they must remain on the plant for longer than deciduous leaves, and that means they must be more robust. They are thicker and have a thick, waxy cuticle.

The broad-leaved plants that are adapted to climates with a long, hot, dry season have taken these characteristics further. They have leaves described as sclerophyllous, from the Greek *sklēros*, meaning hard, and *phyllon*, meaning leaf, or coriaceous, which means leathery. Sclerophyllous leaves are small, thick, and closely spaced—the internodes between leaves are short. Many of the leaf cells have walls containing lignin, and the outer cuticle, or skin, is very tough. There is often a layer of lignified cells immediately below the epidermis that forms a hypodermis, or toughened inner layer. The leaf veins are close together and the cells around them are also fortified. The extensive strengthening allows the leaves to retain their shape in times of drought, when their tissues lose moisture. No matter how dry they are, the plants do not wilt.

Some plants have sclerophyllous leaves that hide their stomata in small depressions on the underside with cuticle extending part of the way over them and sheltering them from direct contact with the air. This reduces moisture losses by transpiration.

Sclerophyllous vegetation occurs in nutrient-poor soils and bogs, but it is predominantly associated with Mediterranean climates, though similar climates occur in other parts of the world, especially in Australia.

Many Mediterranean sclerophyllous plants are cultivated. These include the olive (*Olea europeae*), rosemary (*Rosmarinus officinalis*), sage (*Salvia officinalis*), thyme (*Thymus vulgaris*), and oregano (*Origanum* species).

C3, C4, and CAM photosynthesis

Photosynthesis is the sequence of chemical reactions by which green plants synthesize sugars from carbon dioxide and water, using sunlight as a source of energy. In cells just below the surface of leaves, chlorophyll *a*, which is blue-green, chlorophyll *b*, which is yellow-green, and carotenoids, which are yellow and orange, capture photons of light, each colour absorbing light at a particular wavelength. The energy of each photon excites an electron in the absorbing molecule. The electron then travels from molecule to molecule along an electron transport chain, its energy driving the series of photosynthetic reactions. The overall reaction can be summarized as:

$$6CO_2 + 6H_2O + \text{light energy} \rightarrow C_6H_{12}O_6 + 6O_2\uparrow$$

$C_6H_{12}O_6$ is glucose, which represents the more complex sugars produced by photosynthesis, and the arrow pointing upward indicates that oxygen is released into the air.

Obviously, it is not quite so simple. Photosynthesis proceeds in two stages, the first dependent on light and the second light-independent. During the

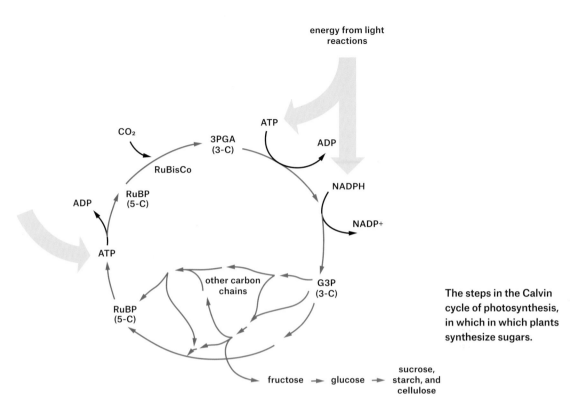

energy from light
reactions

CO₂

3PGA
(3-C)

ATP

ADP

RuBisCo

NADPH

RuBP
(5-C)

ADP

NADP+

ATP

other carbon
chains

G3P
(3-C)

RuBP
(5-C)

fructose → glucose → sucrose,
starch, and
cellulose

The steps in the Calvin
cycle of photosynthesis,
in which in which plants
synthesize sugars.

light-dependent stage, water molecules are split into hydrogen and oxygen. The oxygen is released into the air and the hydrogen is attached to NADP (nicotinamide adenine dinucleotide phosphate), changing it to NADPH. Some of the excited electrons attach phosphate groups to adenosine diphosphate (ADP), changing it to adenosine triphosphate (ATP).

The light-independent stage is also known as the Calvin, Calvin-Benson, or Calvin-Benson-Bassham cycle, after Melvin Calvin, Andrew Benson, and James Bassham, who elucidated its steps. The cycle begins when a molecule of carbon dioxide is attached to a molecule of ribulose biphosphate (RuBP), a sugar containing five carbon atoms, the reaction being catalyzed by the enzyme RuBP carboxylase, or RuBisCo. RuBP, with six carbon atoms, is unstable and divides into two molecules of 3-phosphoglycerate (3PGA), each with three carbon atoms. ATP then donates one phosphate group to the 3PGA, changing it to 1,3-diphosphoglycerate, and NADPH donates two electrons, reducing it to glyceraldehyde 3-phosphate (G3P). Three turns of the cycle produce six molecules of G3P. One G3P leaves the cycle and the remaining five are converted to RuBP. G3P is the primary product of the cycle and, as the diagram above shows,

All brassicas, plants such as these, are C3 plants.

Corn (maize, *Zea mays*) is a C4 grass.

further reactions then convert it to a variety of other carbohydrates.

This is the basic and most widespread pathway of photosynthesis, and the earliest to evolve. It is known as the C3 pathway because its first product, 3PGA, has three carbon atoms. But there is a problem. The key enzyme, RuBisCo, has an equal affinity for oxygen and carbon dioxide and will attach itself to whichever is more abundant, so when carbon dioxide levels are low the C3 pathway becomes inefficient. An alternative has evolved independently approximately 50 times. It first appeared during the Oligocene, about 30 million years ago, when atmospheric carbon dioxide levels were low, is most widespread in tropical and subtropical grasses, and it works best in warm climates with a summer rainy season and high light intensity. It is known as the C4 pathway. Sugar cane and maize (corn) are among the C4 plants.

The C4 pathway begins in mesophyll and bundle sheath cells below the leaf surface. The mesophyll cells are fairly loosely arranged and surround the bundle sheath cells, which are packed more tightly around the leaf veins. In the first step of the pathway carbon dioxide is attached to phosphoenolpyruvate (PEP) with the help of a different enzyme, PEP carboxylase, to produce oxaloacetate (OAA), containing four carbon atoms, which is then converted to another four-carbon compound, usually malate, after which it enters the Calvin cycle. PEP carboxylase has no affinity for oxygen and a higher affinity for carbon dioxide than RuBisCo has. The diagram opposite shows the steps in the process.

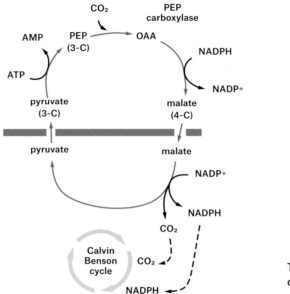

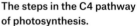

The steps in the C4 pathway of photosynthesis.

Efficient though the C4 pathway is, plants using it must still open their stomata to allow the exchange of gases, and opening the stomata results in transpiration. Plants that grow in hot deserts cannot afford to lose moisture in this way, and they have evolved a strategy for reducing transpiration. The method was first described in stonecrops (*Sedum* species), houseleeks (*Sempervivum* species), and other members of the family Crassulaceae, and it involves an acid, so it is known as Crassulacean acid metabolism (CAM), although it occurs in members of at least 17 other plant families.

Most plants open their stomata by day, because that is when photosynthesis takes place and they need to absorb carbon dioxide and release oxygen. CAM plants, on the other hand, open their stomata at night, when the air is cool and transpiration is at a minimum. They exchange gases but photosynthesis is impossible, so the carbon dioxide passes through the first stages of the C4 pathway and is then stored as malic and other organic acids in vacuoles in mesophyll cells. During the day these acids give up their carbon dioxide and photosynthesis takes place, but the leaf stomata remain firmly closed. The diagram on page 245 shows the steps in CAM photosynthesis.

Photosynthesis and photorespiration

A carboxylase is an enzyme that catalyzes reactions which incorporate carbon atoms, and an enzyme that oxidizes a substrate is an oxidase. RuBisCo is both

The pineapple (*Ananas comosus*), a tropical perennial, is a CAM plant.

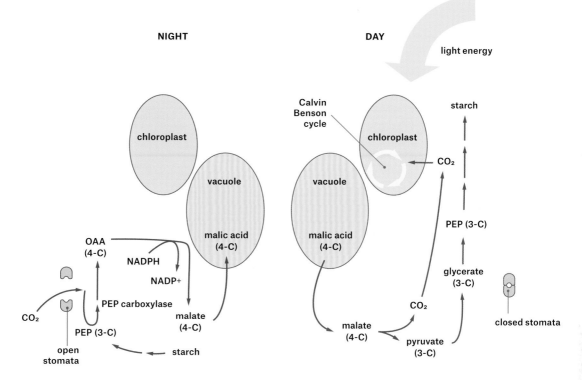

The steps in the CAM pathway of photosynthesis.

and attaches itself to either carbon dioxide or oxygen, depending which is the more abundant. If carbon dioxide is in short supply in the photosynthesizing cells, therefore, RuBisCo acts as an oxidase and instead of photosynthesis an alternative sequence of reactions occurs, known as photorespiration.

True respiration is the opposite of photosynthesis. It is the process in which oxygen is absorbed, carbon is oxidized with a release of energy, and carbon dioxide is released. Photorespiration is similar, but it releases no energy. The diagram on page 246 shows the steps involved in the photorespiration and Calvin cycles.

Photorespiration begins with the oxidation of RuBP (ribulose biphosphate), catalyzed by RuBisCo. RuBP breaks into one molecule of 3-phosphoglycerate (3PGlycerate) and two molecules of phosphoglycolate. The phosphoglycerate is converted to glycolic acid, enters an organelle called a peroxisome, and is converted to glycine. The glycine is transported to a mitochondrion where it is converted to serine. The serine takes part in other reactions that release carbon dioxide and ammonia (NH_3). Energy for the reactions is obtained by converting ATP to ADP.

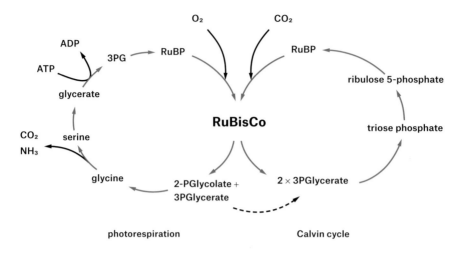

RuBisCo has an affinity for oxygen as well as for carbon dioxide. When it attaches to oxygen it triggers reactions in which oxygen is absorbed and carbon dioxide released to the air, but although this resembles respiration it releases no energy.

Unlike photosynthesis, therefore, photorespiration utilizes ATP rather than adding phosphate to ADP to form ATP as a means of storing energy. It produces no useful metabolic product and removes carbon from the plant. It is of no benefit to the plant whatever. It may have evolved at a time when the air contained much more carbon dioxide than it does today, so that RuBisCo would have acted as a carboxylase most of the time and photorespiration would have been insignificant, but today it reduces the efficiency of photosynthesis in C3 plants by about 25%.

Photoperiodism

I wake up the same time very morning, get hungry in the middle of the day, and grow sleepy when it's getting near my bedtime. At least to this extent my life follows a routine as though there's a clock inside me switching things on and off. And, of course, there is. Every animal and plant has such a clock controlling a daily or circadian rhythm (from Latin *circa*, about, and *dies*, day).

The circadian rhythm is not precise, however. It needs external cues to reset it, just like a watch that doesn't keep good time, and it is the solar clock that gives us day and night which provides the needed prompts. In many species of flowering plants the mechanism regulating the circadian rhythm blends with

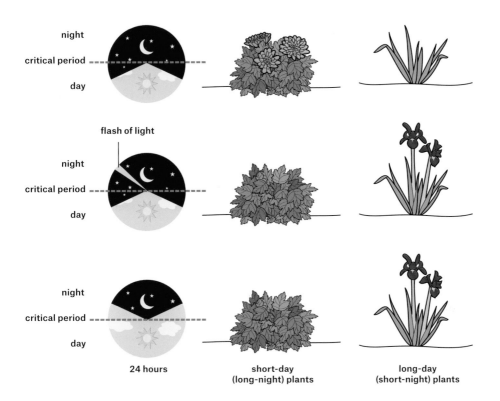

| | night | short-day (long-night) plants | long-day (short-night) plants |

night
critical period
day

flash of light

night
critical period
day

night
critical period
day

24 hours · short-day (long-night) plants · long-day (short-night) plants

Day-neutral plants flower irrespective of changes in the hours of daylight and darkness. Short-day plants flower when the night exceeds a critical length. A flash of red light during the night will prevent flowering, but a subsequent flash of far-red light will reverse the effect. Long-day plants flower when the length of night falls below a critical value. A flash of light during the night does not inhibit flowering.

another system to ensure flowers open at the time of year that is best for the formation of fruit and release of seeds for that species. Species fall into three groups. For some plants, a change in the length of night stimulates them to produce flowers. Other plants are insensitive to changing night length. Because we tend to sleep at night, this variation in plants was first noticed in daylight and they were thought to be responding to changes in the hours of daylight, so they came to be described as short-day, long-day, and day-neutral. In truth, though, it is the hours of darkness that provide the trigger. The phenomenon is called photoperiodism. The illustration above shows the difference between short-day (long-night) and long-day (short-night) plants.

Phytochrome is the substance mediating this response. It comprises two identical protein molecules joined (conjugated) to a non-protein molecule that absorbs light. Plants synthesize five varieties of phytochrome, two of which are involved in photoperiodism. One of these absorbs red light (wavelength about 660 nanometres) and is known as Pr, and the other absorbs far-red light (wavelength about 730 nm) and is known as Pfr. When Pr absorbs red light it changes to Pfr and when Pfr absorbs far-red light it changes to Pr. By day the daylight contains more red light than far-red, so during the day Pr is steadily changing to Pfr. At night the Pfr spontaneously changes back to Pr and it is Pr that is needed to switch on the signal for the plant to flower.

In short-day plants, however, a brief flash of red light during the night, even if all the Pfr has already been changed back to Pr, is sufficient to inhibit flowering. Clearly, therefore, an additional mechanism must be involved, and that is where the circadian rhythm comes in. The circadian rhythm uses phytochromes that absorb red light and cryptochromes that absorb blue light and phytochromes A and B that mediate the degradation of a transcription factor of a gene that switches on another gene, which triggers the formation of flower buds. If the period of darkness is too long, not enough of the transcription factor survives to activate the flowering gene in long-day plants. In short-day plants biologists suspect that instead of switching the flowering gene on, the transcription factor inhibits it.

Rudbeckia, lettuce, and spinach are long-day plants, which is why lettuce and spinach are prone to bolting in summer. *Poinsettia*, Christmas cactus, strawberry, and chrysanthemums are short-day plants and will not flower if the nights are less than 12 hours long. Tomatoes, petunias, asters, sunflowers, and cucumbers are day-neutral plants.

Coping with salt

A plant that tolerates salt is called a halophyte and you are most likely to find it in a salt marsh or other estuarine environment. Conditions there can be extreme and, as well as salt, halophytes must survive extreme hot and cold temperatures, large variations in the availability of water, including waterlogging, and wide variations in salinity. *Atriplex* species (saltbush), *Salicornia* species (glasswort, marsh samphire), *Distichlis spicata* (saltgrass), and *Spartina alterniflora* (smooth cordgrass) are halophytes.

Each group of plants occupies a particular zone of such an intertidal habitat, depending on the frequency and length of time it is immersed in salt water. The diagram opposite shows the pattern that results, extending from dry land

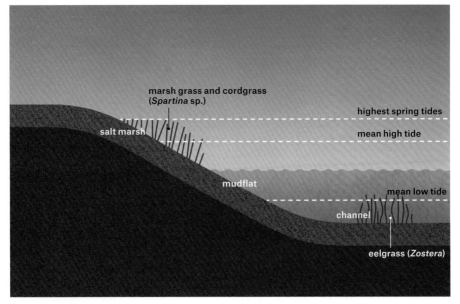

The halophytic plants that occupy zones of a salt marsh extending from dry land to the low-tide line.

above the level of the highest spring tides, where plants are never immersed but are frequently splashed by salt spray, to the mean low-tide line, where plants are submerged most of the time.

Salt dehydrates cells. When the solution in the space between cells is stronger than that inside cells, water moves out of the cells by osmosis. That is why few plants are able to survive in salt or even brackish water. Those that do, however, have fewer competitors for the resources available.

High salinity inhibits seed germination. Many species of *Atriplex* carry their seeds in bracts that contain high levels of salt. The seeds cannot germinate under these conditions, but heavy rain leaches the salt from the bracts and also dilutes the salt in the ground below the plant to a level in which the seeds will germinate. The seeds absorb the fresh rainwater, swell and grow heavier, fall to the ground, and germinate in the seedbed the rain has prepared.

Some halophytes cope with high salt levels by isolating the salt in their tissues and accumulating it. Others exclude salt, thereby avoiding it entirely.

Salicornia rubra (red swampfire) is a halophyte, thriving in saline environments.

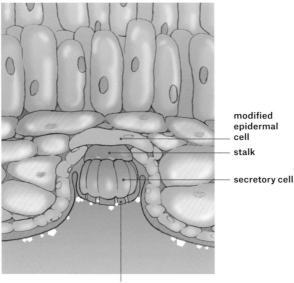

modified
epidermal
cell

stalk

secretory cell

cuticle

A typical salt gland in a dicot halophyte comprises a modified epidermal cell in which salt collects and from where it passes through a stalk to a secretory cell from which it is excreted. The cuticle (dark green layer) is perforated above the salt gland to allow salt to leave.

Atriplex, *Spartina alterniflora*, and *Distichlis spicata* accumulate salt in salt bladders on their leaf surfaces. These are modified hairs, or trichomes, made from pairs of cells that are often ten times larger than cells of the epidermis. Most halophytic grasses have them and their density increases with increasing salt concentration. When the bladders are full they burst. The saline solution flows over the leaves, the salt crystallizes in dry air, and the crystals fall or blow away, or wash away with the next fall of rain. This strategy is widespread in halophytic goosefoots (Amaranthaceae).

Salt glands are somewhat similar and occur in some mangroves and salt grasses. In monocots they comprise two cells and in dicots they are more complex, typically consisting of a modified epidermal cell in which salt collects and from where salt passes through a stalk cell to a secretory cell from which it is excreted. Cuticle covering the gland is perforated to allow the salt to leave. The illustration above shows the structure of a salt gland.

Other accumulators lack specialized salt bladders or glands, but instead transport salt, as sodium and chloride ions, into the cell vacuole. Its increasing salt concentration causes the vacuole to absorb water, and the vacuole expands. The cell remains turgid because it contains the osmotic loss of water. The water remains inside the cell. Halophytes that deal with salt in this way are succulents, such as *Salicornia* species.

Mangrove forest forms monocultures along many tropical coasts, where its roots trap sediment, slowly extending the shoreline seaward.

There are several ways plants can exclude salt. In some, the semipermeable membranes in the outer layer of the root epidermis are modified to block the passage of sodium and chloride ions. Waxy strips in the inner root epidermis perform a similar function. More commonly, plants accumulate salt in their leaves until the concentration exceeds a certain threshold in the petiole. The petiole then detaches from the plant, taking the leaf with it.

Plants and their environment

The climate, which is weather averaged over many years, determines most of the physical parameters within which plants and animals must live. Living organisms that are just slightly better than those around them at exploiting their environment within those parameters produce more offspring. In a word, they thrive.

And those that thrive become more abundant. That is how evolution works to produce new species and also communities of those plants and animals that are best suited to the circumstances in which they occur.

Having set the parameters by considering the way physical forces produce our day-to-day weather and the way weather produces climates, and having thought about some of the ways plants are adapted to the climates where they occur, it is time to consider the patterns that result. Were you to visit a particular environment, that is the product of climate, such as prairie or Old World desert, what types of plants should you expect to find there?

Typical plants of Old World deserts

Think of the Sahara and the image that springs to mind is of high sand dunes beneath a pale blue sky and searing heat. Surely, no plant could survive in such a hostile place, nor could its roots find purchase in the shifting sand. Yet single palm trees, the most typical of all desert plants, sometimes rise from the dunes in dramatic isolation, indicating the location of water deep below the desert sands.

Desert palm trees have deep roots that can reach the groundwater underlying large areas of the Sahara and the date palm (*Phoenix dactylifera*) is an important crop plant that has been cultivated since about 4000 BCE. The tree grows to about 20 metres. Traditionally, growers ensured pollination by cutting off bunches of male flowers and hanging them among bunches of female flowers. Today they are more likely to split open the mature female spathe with a metal implement at the end of a long aluminium tube or bamboo pole and inject dried pollen with a puff of air. Each female tree bears several bunches each of more than 1000 dates.

Figs (*Ficus carica*) are now grown wherever the climate is warm enough, but they were cultivated in ancient Egypt and possibly even earlier in Jericho. The cultivated mulberry (*Morus nigra*) and almond (*Prunus dulcis*) are natives of the desert climates of the Middle East.

Desert euphorbias are succulents and the Old World equivalent of the cacti. Some are the size of small trees, the candelabra tree (*Euphorbia candelabrum*) reaching about 11 metres, with multiple branches emerging about 3 metres above the ground and rising almost vertically—like candles in a candelabra. Ice plants and living stones (*Lithops*) are natives of the Kalahari and Namib Deserts.

The strangest of all desert plants, and perhaps of all plants anywhere, is welwitschia (*Welwitschia mirabilis*), which is also a native of the Namib Desert,

Welwitschia mirabilis grows in and close to the Namib Desert. It produces only two leaves, which continue to grow throughout the centuries of the plant's life, becoming very frayed at the ends.

and is found only there and in woodland on nearby mountains. The plant has a short, thick stem, rising to about 50 centimetres above ground and growing from a long taproot. At the top the stem divides into two lobes, each of which produces a single, strap-like, leathery leaf. The two leaves grow at a rate of about 13 centimetres a year and continue to grow throughout the life of the plant. As they grow longer the leaves fray at the ends, so they look like thongs and lose their ends, but by then the leaves are up to 4 metres long. The plant lives for up to 600 years and some specimens may be more than 1500 years old. It is grown as a garden ornamental—or more accurately, perhaps, as a conversation piece—by gardeners in dry climates. Like all desert plants, it is vulnerable to fungal diseases if it is too moist.

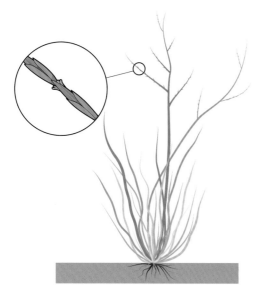

Tumble grass (*Schedonnardus paniculatus*) produces flowers in branched, terminal panicles. Once it has set seed the panicles detach from the root and curl into a ball as they wither.

Typical plants of New World deserts

Cacti are natives of American deserts, but cacti, being succulents, have a well-fed, comfortable look despite their spines. They are attractive plants and are cultivated all over the world for that reason. The plant that evokes like no other the desolation and wood-shrinking aridity gripping a Western homestead or small town is not the luxuriant cactus but the forlorn tumbleweed that appears in so many movies.

Tumbleweed is not a single plant species, but a solution to a problem that several species have adopted, in the Old World as well as the New. The species most often featured in westerns is tumble grass (*Schedonnardus paniculatus*), a true grass (Poaceae), and the problem it solves is that of distributing its seeds.

Tumble grass is a perennial with tangled stems up to about 50 centimetres long and blades up to about 10 centimetres long, most growing from the base. The inflorescences are branched, terminal panicles up to 50 centimetres long. Tumble grass grows in low tufts. The illustration above shows the growing plant with its long panicles. When the plant has set seed, the panicles detach from the body of the plant and, as they wither, curl into a tangled ball with the seeds on the inside. Then the wind blows them about, scattering seeds as they go. Where moisture and nutrients are scarce, plants that simply allow their seeds to fall will have to compete for resources with their own offspring. Scattering them over a wide area avoids competition and gives the seeds a better chance of germinating.

Tumble grass, however, is rather insignificant except when it is distributing its seed. The plant you are most likely to see is the creosote bush or greasewood (*Larrea divaricata*). Its success as a desert plant is beyond question. In the Mojave Desert there is a ring of creosote bushes known as King Clone, which have been radiocarbon dated as 11,700 years old. Chemicals secreted from the roots of creosote bushes inhibit the growth of other plants except for burro bush or bur-sage (*Ambrosia dumosa*), the two often occurring together. Creosote bushes grow to about 3 metres tall. The plant is evergreen with resinous leaves that smell of creosote. A liquor made by steeping the twigs has medicinal properties. The creosote bush can survive without rain for more than a year, shedding all its leaves and its leaf buds becoming dormant. It reproduces by suckering, so groups of adjacent bushes are clones, and except in dry years they also produce yellow flowers and set seed. Gardeners who cultivate them sometimes use the unopened flower buds like capers.

Most agaves also originate in the American deserts and are cultivated worldwide for their long, succulent leaves, many with prickles along the edges. The century plant or American aloe (*Agave americana*) is a popular garden plant, named for the mistaken belief that it flowers only once in a century. In fact, it flowers at intervals of 10–20 years.

Yuccas are close relatives of the agaves. Most yuccas are pollinated by *Tegeticula* species of nocturnal moths, and many yucca flowers last only one night. The most spectacular yucca is probably the Joshua tree (*Yucca brevifolia*).

All *Banksia* species are endemic to Australia except for *B. dentata,* which occurs in northern Australia and the islands to the north. There are no species that occur naturally in both the west and east of Australia.

Australian plants

Banksias are Australian and all of the 173 species are endemic, except for *Banksia dentata* (tropical banksia), which occurs across northern Australia and also in Irian Jaya, Papua New Guinea, and the Aru Islands. The remaining species occur on the western and eastern sides of the country, with no species common to both. The map shows the distribution of banksias.

Banksias belong to the family Proteaceae, as do the five species of *Telopea* known as waratahs. Waratahs are widely cultivated for their red, pink, or yellow flowers and *T. speciosissima* is the state flower of New South Wales. Bottlebrushes, 40 species of *Callistemon*, occur mostly in the east and

Eucalyptus regnans (giant or mountain ash, among many other common names), native to Australia, the world's tallest angiosperm.

River red or red river gum (*Eucalyptus camadulensis*) grows widely
in Australia, especially beside rivers.

southwest of Australia, growing in damp or wet ground. Several species are
cultivated for their bottlebrush-shaped flowers. The plants are shrubs up to 4
metres tall. Kangaroo paws, 11 species of *Anigozanthos*, are native to the southwest of Western Australia but are popular garden plants and are grown commercially in some countries.

Australia's largest genus of vascular plants is *Acacia*, comprising 948 species of trees and shrubs known as wattles. These are now believed to be distinct from similar plants that grow in other parts of the world. Wattles occur
throughout Australia in a variety of habitats, but they are especially common
in the arid, semi-arid, and dry subtropical parts of the country. The golden wat-
tle (*A. pycnantha*) is Australia's national flower. Wattles are grown widely as
garden plants.

The world's tallest flowering plant and one of the tallest trees, native to Victoria and Tasmania, is *Eucalyptus regnans*, known as the giant ash, mountain

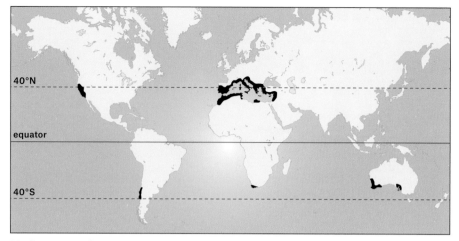

Mediterranean climates occur in both hemispheres on the western side of continents between latitudes 30° and 40°.

ash, Victorian ash, swamp gum, stringy gum, and Tasmanian oak. Individuals have grown to more than 114 metres tall. It is evergreen and grows in mountainous areas with an annual rainfall of more than 1200 mm (47 in.).

There are more than 700 species of *Eucalyptus* and most are native to Australia where they are the predominant forest trees. They are often called gum trees because of their sap. Many species are grown throughout the world for ornament or commercially for timber.

Eucalyptus leaves release volatile terpene compounds, which give the trees their pleasant smell, and the higher the air temperature the more they produce. Terpenes react with other atmospheric gases to produce the blue, smog-like haze that gives the Blue Mountains their name. Eucalyptus oil is highly volatile and flammable, and once eucalyptus trees catch fire they burn rapidly and fiercely, and sometimes explode. The abundant leaf litter in eucalyptus forests is also very flammable and accumulates because it contains phenolic compounds that inhibit its breakdown by fungi. Consequently, eucalyptus forests and plantations can be sites of catastrophic fires.

Plants of Mediterranean climates

The subtropical highs are permanent anticyclones formed by air subsiding on the poleward sides of the Hadley cells, and centred over the oceans between latitudes 30° and 40° in both hemispheres. The circulation of air associated with the subtropical highs produces a distinctive type of climate on the

Mediterranean maquis vegetation comprises plants adapted to hot, dry summers and mild, wet winters. They have deep roots and small, sclerophyllous leaves.

western sides of North and South America, Eurasia, South Africa, and Australia. In these regions, shown on the map opposite, winters are mild and wet and summers are warm and dry. The largest area with such a climate occupies the coastal regions around the Mediterranean, so it is described as a Mediterranean climate regardless of where it is found.

The ecosystems produced by a Mediterranean climate are known as maquis and garrigue in France, macchia in Italy, matorral and tornillares in Spain, matorral in Chile, phrygana in Greece, batha in Israel, chaparral in the United States, fynbos in South Africa, and mallee in Australia. Each of these regions supports a different suite of plants, but all of the plants are adapted to similar conditions. They must tolerate the heat and drought of summer, so they have deep roots to reach a low water table, and small, sclerophyllous leaves that keep their stomata closed when it is necessary to conserve moisture. Some

Chaparral is the Mediterranean-type vegetation of the western United States.

plants produce large leaves in winter and small ones in summer. Annual plants release their seeds in summer to wait for the autumn rains to germinate.

The vegetation is also adapted to fires, which are frequent in summer. Some have very dense wood or fireproof bark, allowing them to survive fire. Others are geophytes—plants with underground storage organs such as rhizomes, tubers, corms, or bulbs that give rise to buds from which new shoots emerge when conditions above ground are favourable. These include *Arbutus*, *Erica*, *Juniperus*, and *Pistacia* species. Many plants of the chaparral and most of those of the Chilean matorral produce new shoots from the base following fire.

There are also serotinous plants, which depend on fire. Serotiny is the habit of some trees of retaining their seeds in pods or cones, sometimes for several years, until a major event causes them to be released. That event is usually the heat from a fire. The seeds fall on to soil enriched by ash where they germinate with little competition. Many *Eucalyptus* trees are serotinous, as are jack pine

The dragon tree (*Dracaena draco*) grows in the arid regions of the Canary Islands and adjacent Africa. It is widely cultivated, but becoming rare in the wild.

The Mallee comprises scrub vegetation that covers large areas of Australia. Eucalypts are the dominant trees. This is *Eucalyptus loxophleba* (York gum).

(*Pinus banksiana*), lodgepole pine (*P. contorta*), and other *Pinus* species.

Mediterranean vegetation comprises forests, although most of those have been cleared around the Mediterranean itself, shrublands, and grasslands. Typical Mediterranean plants include the olive (*Olea europaea*), holm oak (*Quercus ilex*), bay (*Laurus nobilis*), juniper (*Juniperus* species), pistachio (*Pistacia* species), and carob (*Ceratonia siliqua*).

Ceanothus species (California lilac) are the most widely cultivated of the chaparral plants. Chamise (*Adenostoma fasciculatum*), scrub oak (*Quercus berberidifolia*), and silktassel (*Garrya flavescens* and *G. veatchii*) are other common chaparral plants.

The plants of the Chilean matorral are less familiar, but they include the Chilean wine palm (*Jubaea chilensis*), said to be the hardiest of all palms, tolerating temperatures below –12°C (10°F).

The king protea (*Protea cynaroides*) is the most famous plant of the fynbos and South Africa's national flower.

Eucalypts, most no more than 6 metres tall, are the dominant trees of the Australian mallee. There are also wattles (*Acacia* species), banksias, and *Melaleuca* species, which are trees and shrubs related to myrtles (Myrtaceae), many of them popular with gardeners in the tropics.

The South African fynbos is one of the world's richest ecosystems and the source of many garden plants. These include *Protea* species, heaths (*Erica* species), asters, irises, and the grass-like *Restio* species, and many more.

barren
rocks

mosses and lichens

cushion plants

open grassland

alpine meadows

(dwarf) bushes

coniferous forest

deciduous forest

The vegetation on a mountainside forms distinct altitudinal zones. The altitude of each zone varies with latitude and the composition of each zone varies geographically.

Mountain plants

As you climb a mountain the temperature falls and the vegetation changes with altitude. The plants are related to those growing on the plains at the foot of the mountain, but they are adapted to the conditions in which they grow. On any mountainside, therefore, the plants at each elevation will be similar through adaptation, but not necessarily of the same species.

The illustration above shows the vegetation zones on a typical mountain-side—in fact in the European Alps—but the mountain could be anywhere with a temperate climate. Deciduous forest cloaks the foot of the mountain. Above it, where conditions are too harsh for standard broad-leaved trees, there is coniferous forest and above that there are dwarf specimens of the trees mixed with shrubs. Higher still, there is alpine meadow, rich in colourful flowering herbs. Traditionally, the alpine meadows are where livestock spent the summer after spending the winter in the valleys, tended by herders who lived in mountain huts until autumn, in the practice known as transhumance. Climb higher and the alpine meadow gives way to open grassland and beyond that mosses and lichens replace the grasses at the upper limit of plant growth. The top of the mountain is bare rock and, depending on the height and latitude, covered with snow for part or all of the year.

An alpine meadow, also called alpine tundra, often provides summer pasture for livestock. This meadow is at Garwhal Uttarakhand, in the Indian Himalayas.

The illustration provides no clue as to the altitude where each zone begins and ends, because that depends mainly on the latitude and partly on the aspect, boundaries between zones being higher on the side of the mountain facing the equator than on the cooler side facing away from the equator. Arctic conditions extend all the way to the foot of the mountain in any latitude higher than 70°. Depending on the aspect, in temperate regions, between latitudes 40° and 50°, the transition between deciduous and coniferous forest is at around 2000–3000 metres and the limit for plant growth is at 3000–4000 metres.

Rain forest is likely to flourish at the foot of a tropical mountain. With increasing altitude the forest becomes more open and the trees smaller, and mosses and lichens are more abundant and prominent. For much of the day clouds often shroud tropical mountains above about 1000 metres, producing cloud forest, in which the moisture encourages the prolific growth of epiphytes,

continuous
snow

alpine zone
steppe with tree-like
Asteraceae

subalpine
elfin woodland

montane forest
mossy forest

transitional submontane forest

tropical rain forest

The vegetation zones on a
tropical mountainside, in this
example in the Peruvian Andes,
begins with tropical rain forest
and proceeds through sev-
eral types of woodland before
becoming steppe-like grassland.

especially ferns and mosses. In the mossy forest above the cloud forest, most of
the trees are 10–15 metres tall and their trunks and boughs are festooned with
mosses, lichens, and liverworts. Mosses are especially abundant where mist
envelopes the forest most of the time. The elfin woodland consists of gnarled,
stunted, dwarf trees and beyond the tree line it gives way to steppe-like grassland. The illustration above shows the vegetation zones on the mountainsides
of the Peruvian Andes and the table opposite shows the heights of the zones.

Alpine plants supply the basis for an entire category of gardening and
rockeries are simulations of mountainsides. Apart from the alpines, the most
familiar mountain plants are probably the rhododendrons. There are more
than 1000 species and their classification is complicated. They occur naturally in many parts of Eurasia, North America, and Australia, but the greatest
diversity is found in the Himalayas, and also in the mountains of Korea, Japan,
and Taiwan. Many species were imported to Europe in the nineteenth century
to satisfy the demand of private gardeners and landowners for what were then
exotic shrubs and trees with magnificent flowers. These introductions caused
no harm, with one exception. *Rhododendron ponticum*, often called wild rhododendron, has become very invasive. It formerly provided rootstock for cultivated rhododendrons and was planted in woodlands to provide winter cover

PLANTS AND THEIR ENVIRONMENT

It is the year-round warm temperatures and high precipitation that make tropical rain forest so lush.

TROPICAL MOUNTAINSIDE VEGETATION ZONES

Altitude (metres above sea level)	Vegetation zone
0 to 1000	tropical rain forest
1000 to 2000	cloud forest
2000 to 3000	mossy forest
3000 to 4000	elfin woodland
4000 to 5000	steppe-like grassland
5000 to 6000	upper limit of plant growth
above 6000	permanent snow, bare rock

Rhododendron, growing in Himalayas, produces vast splashes of colour. Their large, showy flowers led nurseries to send their plant collectors in search of species that would thrive in European and American parks and gardens.

for pheasants. It is difficult to remove, although it can be controlled. *Rhododendron ponticum* is also susceptible to infection by *Phytophthora ramorum*, the pathogen that causes sudden oak death, which increases the urgency of the efforts to bring rhododendron under control.

Plants of grasslands

Grasses can thrive wherever the climate is too dry for most woody plants, so temperate grasslands occur naturally on the plains in the interior of continents and tropical grasslands between the forests and the edge of the desert. The map opposite shows the location of the world's grasslands. Temperate grasslands are known as prairie in North America, pampas in South America, steppe in Eurasia, and veld in South Africa. Tropical grasslands are called savannah.

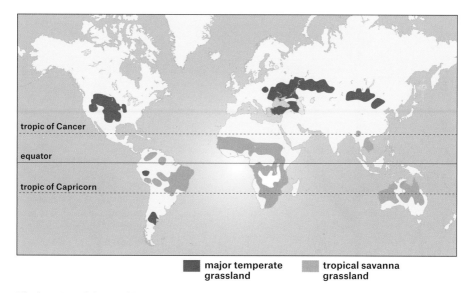

| major temperate grassland | tropical savanna grassland |

The location of the world's tropical and temperate grasslands.

Climate is only one of three factors favouring grasslands. The others are traditional methods of land management, and herds of herbivorous mammals. For thousands of years prior to the invention of agriculture, hunters pursued game across the grasslands and they discovered they could use fire to drive the herds into traps. Fire destroyed trees and shrubs. These would recover from seeds dormant in the soil, but grasses recovered faster, so with each fire the grasses increased their hold. Some of the game animals fed by browsing on shrubs and low tree branches, others grazed. The grazers flourished with the increasingly abundant grass and as they grazed they nibbled and trampled the seedlings of woody plants, little by little destroying most of them with the result that in the end the grasses triumphed.

Grazing kills most plants, but not grass. As we all know, mowing the grass—equivalent to grazing—encourages it. And that is because of the way all grass plants are made. The illustration on page 270 shows a grass plant. The stem, called the culm, grows from meristem cells at the top of each node. If a herbivore or mower removes the upper part of the culm, it grows back from a node lower down. So you can't kill grass by grazing or mowing it. You can't kill it by trampling, either. That is because if the culm is flattened its growth hormones accumulate on the underside, causing the meristem cells on that side to grow faster than those on the upper side. So the flattened plant rises once more from a node at ground level.

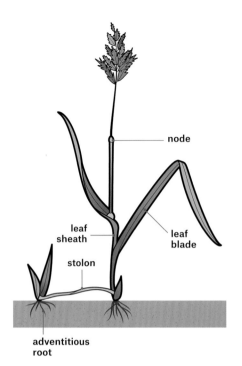

node

leaf
sheath

leaf
blade

stolon

adventitious
root

All grass plants share certain features, including their ability to respond to cutting by continuing to grow from the node beneath the cut, to respond to hot, dry air by curling their blades so the stomata are sheltered, their production of fibrous root mats, and in many species the ability to spread by means of rhizomes or stolons.

There are variants to this plan. Most grasses have a single culm, but some species produce many culms from the base. This is called tillering. It defines tussock grasses and is also typical of wheat, barley, oats, and rye, the grasses that feed us. Some tropical grasses, especially bamboos, produce branches from nodes on the upper part of the culm.

Grass roots form a dense, wide mat that absorbs water and nutrients from just below the surface, taking quick advantage of rain before it evaporates, but many prairie grasses also have deep roots. The roots also anchor the plant securely. Some grasses are annuals, growing from seed each season, but many are perennials and expand by rhizomes or stolons. Meristem cells on the bottom of nodes along rhizomes and stolons give rise to adventitious roots that serve the plants growing upward from the nodes.

Prairie grasses vary in height according to the amount of water available to them. Short-grass prairie grows where rain moistens only the upper 30 centimetres of soil and the ground is often dry. This type of prairie occurs in the Midwest, from southern Alberta and Saskatchewan to Texas and New Mexico. These are the rangelands, with most grasses no more than 20 centimetres tall. Tall-grass prairie grows farther east. Big bluestem (*Andropogon gerardii*) is the tallest of the grasses. It grows in tussocks with culms from 60 centimetres to 3 metres tall and leaves up to 30 centimetres long. Its roots penetrate 2 metres below the surface. Indian grass, also known as wood grass, yellow Indian grass, bushy bluestem, and wild oat grass (*Sorghastrum nutans*) grows to more than 2 metres tall, usually in large clumps.

Silver pampas grass, also called Uruguayan pampas grass (*Cortaderia selloana*) is the most widely cultivated of the pampas tall grasses. It grows in large tussocks with sharp-edged leaves up to 2 metres long and a culm growing to nearly 4 metres, topped by inflorescences forming feathery plumes. The pampas support short and tall grasses and, as with the prairies, the tall grasses grow on the moister eastern side of South America. The climate there is wetter than that of eastern North America, however, and the grasses are taller. When the first Spanish explorers came ashore at the mouth of the La Plata River in the sixteenth century they gazed across a waving sea of silver pampas grass so tall they had to stand on the backs of their horses to see over it.

Tallgrass prairie grows on the moister, eastern side of North America. Farmland has replaced most of this type of prairie.

Some plant geographers describe as prairie all temperate grasslands with a moist climate and those with a dry climate they call steppe. By this definition parts of the South American grasslands, found mainly in central Argentina, are steppe. There the dominant plants are needle grasses, also known as feather grasses and spear grasses (*Stipa* species) that grow in dense tussocks up to 60 centimetres tall. They have very narrow leaves that give the tussocks a feathery appearance.

The largest area of steppe extends for 9000 kilometres from the Danube to Mongolia and western China and from the edge of the Sahara northward to the taiga. At least, that is the area they once covered. Today much of the steppe has been converted to agriculture and forests have been planted across the northern steppe. Needle grasses are abundant, in places extending as far as the eye can see. Lessing's (*Stipa lessingiana*) and Ukrainian feather grass (*S. ucrainica*) are predominant in the west and June grass (*Koeleria macrantha*) grows in the east, where the climate is drier. In the north, forest steppe borders and merges into the taiga, the trees becoming more widely scattered farther south. To its south the meadow steppe, with a moister climate, supports sedges (*Carex* species) and needle grasses. Dry steppe, to the south of the meadow steppe has shorter needle grasses.

Red oat grass (*Themeda triandra*), up to 90 centimetres tall, is the most widespread grass of the South African veld, also called the grassveld. The

The steppe grasslands extend from the Danube to Mongolia. This landscape is in Kyrgyzstan.

The African savannah supports tropical grasses with scattered thorn trees that are renowned for their spreading, layered foliage.

veld is farmed. It supplies much of the nation's beef, dairy products, and wool, and it also grows arable crops, especially maize (corn), wheat, sorghum, and sunflowers.

Tropical grasslands are known as savannah, a name originally applied to grasslands on some Caribbean islands. Today savannah extends across Africa in two belts, one on each side of the equator, and is also found in Australia, southern Asia, and South America. Red oat grass grows on clay soils where the annual rainfall is 600–1000 mm (23.6–39.3 in.) and also in areas prone to fires, from which the grass recovers quickly. Elephant grass, also called Napier grass and Uganda grass (*Pennisetum purpureum*) grows beside African lakes and rivers. It has leaves with very sharp edges and its culms grow up to 3 metres tall.

The African savannah is also renowned for its thorn trees. Formerly these were placed in the genus *Acacia*, but now *Acacia* is reserved for Australian wattles and the African species have not yet been assigned a generic name. The umbrella thorn is the typical savannah tree. It is umbrella-shaped and up to 15 metres tall.

Thorn trees are scattered because each one needs a minimum area below ground to find water. Their isolation leaves them exposed to herbivores hungry for their young leaves and shoots. The fearsome thorns afford a degree of protection and some thorn trees have swollen thorns at the base of each leaf. Ants hollow out these thorns and live inside them feeding on nectar from nectaries at the base of each petiole and on oils and proteins produced in sausage-shaped Beltian bodies at the leaf tips. Worker ants from each colony defend their territories, fiercely attacking any animal that comes within their reach, and they also cut away any parts of a neighbouring plant that touches their tree.

The boreal forest or taiga extends as a belt across North America and Eurasia and accounts for 29% of the world's forest area.

Boreal forests

Think of a vast area of forest and you might picture the South American tropical forest, seen from an aircraft and stretching from horizon to horizon in every direction. The picture is real, of course, but it springs to mind because the tropical forests have good PR. The world's largest forest is nowhere near the tropics, however. It extends as a belt across North America and Eurasia, roughly between latitudes 50° N and 70° N, and it accounts for about 29% of the world's total forest area. Russians call it the taiga. It is also called the boreal forest. Boreas was the Greek god of the north wind. Winters are long and cold, summers cool and nowhere lasting more than four months, and annual precipitation ranges from 200 to 750 mm (7.9–29.5 in.), much of it falling as snow, but including fog. The climates are of the Dfc, Dwc, Dsc, Dfd, and Dwd types in the Köppen classification. Adapted to average annual temperatures between –5°C (23°F) and 5°C (41°F), the plants begin growing as soon as the daily average temperature is securely above freezing, giving a growing season of up to 150 days in some places but in others of no more than 80 days.

Coniferous trees are the dominant plants. The composition of the forest varies from place to place, but the most common trees are larches (*Larix* species), spruce (*Picea* species), fir (*Abies* species), and pine (*Pinus* species). Spruce and fir are predominant in North America, Scots pine (*P. sylvestris*) in Scandinavia and western Russia, and larch farther east. Dahurian larch (*Larix gmelinii*), native to eastern Siberia and Mongolia, tolerates extreme cold. There are also some broad-leaved trees. Birch (*Betula* species), aspen (*Populus* species), willow (*Salix* species), and rowan (*Sorbus* species) are the most common.

The northern taiga extends toward the limit for tree growth. Travelling north, the trees are increasingly widely scattered and often stunted, and lichens cover much of the open ground. Travel south and the forest becomes denser, with a closed canopy, and in the southernmost part of the taiga the closed conifer forest contains some broad-leaved trees more typical of temperate forest, including maple (*Acer* species) and oak (*Quercus* species).

Flowering herbs flourish where there are clearings in the closed-canopy forest. Fireweed, also called great willowherb and rosebay willowherb (*Chamerion angustifolium*) is prominent. The conifer canopy is dense, however, and casts a deep shade. This restricts the growth of understory trees and ground plants.

Mixed forests

Have you noticed just how many of our folk tales and fairy stories take place in forests or involve journeys into forests? The forests are mysterious places where dangers may lurk and the human inhabitants are poor woodcutters or charcoal burners. Or they are outlaws, driven from their parishes for some, probably trivial, misdemeanour.

The reason is that once upon a time forests covered most of western Europe, with the taiga to the north and to the south of the conifers mixed forests of both coniferous and broad-leaved deciduous trees, and purely broad-leaved deciduous forests to the south of those. These are sometimes called summer forests because the canopy closes only in summer. And it is that seasonality of the canopy that produces the great botanic as well as scenic diversity of these forests, and their beauty.

Apart from the southern tip of South America, temperate broad-leaved deciduous forests are confined to the Northern Hemisphere, because it is only in the Northern Hemisphere that there is dry land where temperatures

Summer or temperate deciduous forest grows where temperatures and precipitation are moderate.

fall below freezing in the coldest month of the year and average more than 10°C (50°F) in the warmest month, providing the conditions such trees require. As well as in western Europe, this type of forest occurs in parts of the Near East and Asia, and in eastern North America, where it is still extensive.

Most of the European forest was cleared centuries ago to provide agricultural land, although nowadays there is widespread planting with native species, mainly for conservation and amenity purposes. Where summer forest is the vegetation naturally favoured by the climate, it re-establishes fairly readily on land that is abandoned, and has done so in large areas of the northeastern United States. There, more than half of the original mixed forest was cleared for farmland and most of the remainder was cut over by logging between about 1650 and 1850. Then the opening up of farmland to the west flooded the eastern states with cheap grain and the bankrupted northeastern farms were abandoned. The forest has returned, but with an altered composition.

Since this type of forest is the natural vegetation, and since there is such a wide diversity of tree species, most of them attractive, it is not surprising that many popular garden trees are native. Those with gardens large enough grow oaks (*Quercus* species), limes (*Tilia* species), ash (*Fraxinus* species), birch (*Betula* species), maple (*Acer* species), and beech (*Fagus* species), as well as introduced species, mainly from North America. These include magnolia (*Magnolia* species), hickory (*Carya* species), tulip tree (*Liriodendron tulipifera*), and sweet gum (*Liquidambar styraciflua*). And, of course, our spring flowers are woodland species.

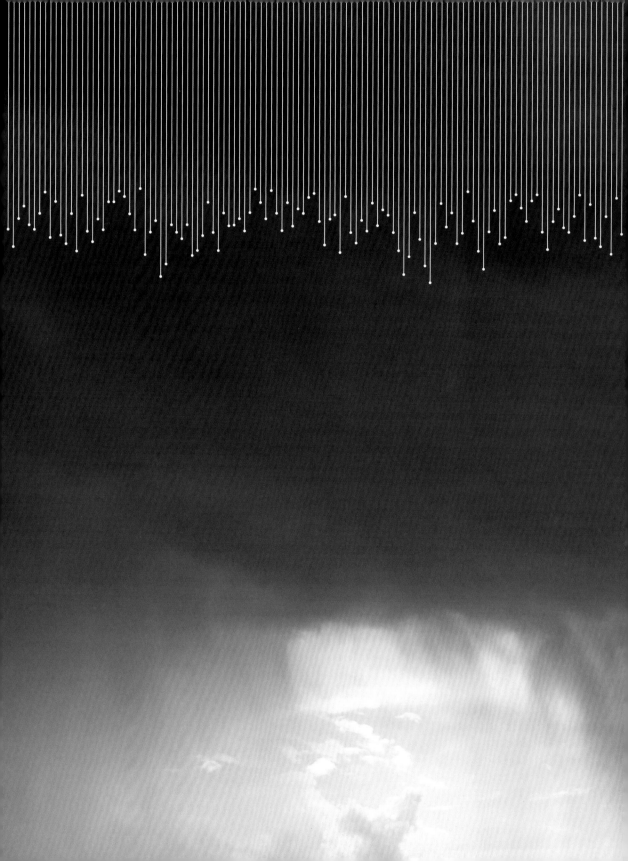

Protecting against harsh weather

We cannot control the weather, no matter how dearly we'd love to. The long history of attempts at weather modification show very clearly that it simply doesn't work. But that doesn't leave us doomed to remain hapless victims of what-ever the capricious elements choose to throw at us. We can't alter the weather, but we can take steps to protect ourselves, and our gardens, against its worst excesses.

Our complaints against the weather are quite simple. It supplies us with too little water so our plants wilt, or too much water so our ground becomes waterlogged and the plants drown. Its winds blow so strongly that they burn plants by dehydrating them. Finally, at least in northern latitudes, it is often too cold for comfort—for us and the plants.

In this final chapter we can consider a few of the ways to deal with bad weather by avoiding it. Over a prolonged period, ordinary, day-to-day weather accumulates to produce a climate. Predicting the weather is difficult, but studying the local climate is not. There are records.

If you have lived in an area for many years, obviously you will be familiar with its climate. You will know how low winter temperatures usually fall and the frequency of severe winters. You will know the directions of the prevailing winds. And you will know the average monthly precipitation and the form in which it falls—rain, snow, hail, or fog. If, on the other hand, you are new to the area, then it may be worthwhile looking up these climatic parameters. If you live in the United Kingdom the basic records are held at regional centres of the Met Office and are accessible from the Met Office website. The records for each station include mean daily maximum and minimum temperature, number of days with air frost, average annual rainfall (snowfall converted to rainfall equivalent), and number of hours of sunshine. The records are continuous for several decades, the length of the run varying from station to station. Lerwick, for instance, begins in 1930, Oxford in 1853. NOAA (National Oceanic and Atmospheric Administration) holds the records for the United States, the Government of Canada holds them for nearly 1500 locations in that country, and no matter where you live you should have no difficulty accessing the data you need online. Failing that, and if you need much more local information, try the library.

Adding water and removing it

It is its mineral composition that determines the ability of a soil to remain moist during a prolonged spell of dry weather. The larger the soil particles and the more irregular their shape, the larger will be the soil pores and the more rapidly water and dissolved nutrients will flow out of the soil. Sand is the extreme example of a coarse-grained soil with poor water retention. Think how quickly a sandy beach dries after the tide recedes. If your soil is sandy it will feel gritty when you rub it between your fingers.

The solution is to spread a layer of absorbent material on the surface and

The water in this ditch is used to irrigate the adjacent fields.

then work it into the upper soil. Composted plant material is the best and most reliable, and an 8-centimetre layer should be sufficient. You can also use grass cuttings, straw, sawdust, wood chippings, and any other soil improver that will help bind the mineral grains together and reduce the pore space. If you use material rich in carbon, such as grass cuttings or wood, you will need to add a nitrogen fertilizer because the bacteria that decompose the carbon-based material will absorb nitrogen from the soil. Composting plant material before using it ensures it has a better carbon to nitrogen ratio. You will need to repeat the treatment each season until the soil contains about 5% of organic matter by volume.

Other soils retain too much water, clay being the extreme example. The same treatment will improve a clay soil, but on no account should you try to improve clay by mixing in sand. When it dries, and it will, the soil will be hard as concrete.

The lines of trees form windbreaks that shelter the crops from the prevailing wind.

If you have an area that is sodden for much of the time you can install drainage to remove surplus water, or you can accept the situation and go the rest of the way by converting the area to a bog. Dig out the area that will become bog to a depth of about 1 metre. If you're making a bog on ground that drains well, line the bottom with pond liner and use a fork to puncture it so it doesn't become a pond. Add gravel, compost, and backfill the material you removed. Then wait for it to become wet enough to introduce bog plants.

Farmers install field drains to remove surplus water and these work just as well in gardens. The drain itself is made from perforated PVC-U pipe. Installed correctly, water flows into and then along the pipe, moving downhill, of course. Alternatively you could hire a contractor to install a mole drain. That is a tunnel cut through the soil at a depth of about 75 centimetres by a bullet-shaped device on the end of a vertical arm. The drain remains open and functional for about ten years, but it needs a big, powerful tractor to cut it.

If you do install drains, think first about where they will discharge the water they collect. You cannot discharge it on to a road, or into a stream, or on to your neighbour's land. You may need to dig a soak-away (or dry well) for it or feed it into the rainwater drain from your house.

Wind shelter and windbreaks

If your garden is exposed to the wind you may need to erect a barrier. This should lie at right angles to the direction of the prevailing wind. The first step, therefore, is to determine that direction. If you install a windbreak in the wrong place, you could make matters worse, funnelling the wind into your plants instead of providing shelter for them.

Information about the winds in your area may be obtainable locally, from your neighbours perhaps, but winds vary over quite short distances so the information you unearth may not be accurate for your location. In any case, it is not difficult to work out for yourself. The word *prevailing* is a clue to the disadvantage in doing that, however. The prevailing wind is the direction from which the wind blows most often and to be of any use, *often* means in the course of the year, so working it out will take you a year and you'll have to take a measurement and plot it once every day. It's quite a commitment but unless you do it properly there's no point in doing it at all.

The task facing you is to make a wind rose. It sounds simple and it is. It just takes a long time. Take a large sheet of paper, robust enough to last the year you'll be working on it. In the centre draw a circle of convenient size, say 10 centimetres in diameter, and mark the circumference with 20-degree divisions, with 0° (N) at the top. You will also need a wind vane and a compass. Mount the wind vane as high as you can, ideally 10 metres above ground level and clear of obstructions. The roof would be a good place.

Every day at the same time, measure the wind direction to the nearest 20°. It is only the direction that matters, not the speed. On your wind rose, measure the wind direction you just recorded from the centre of the circle and draw a line of a convenient length, say 1 centimetre, in that direction. Remember that the wind direction is the direction the wind is blowing from, not the direction in which the air is moving. As the days go by, each time the wind is from a previous direction you lengthen the line on the wind rose in that direction by the standard length (1 cm). Over the year some lines will grow longer than others and eventually one line will be the longest of them all. That is the direction of the prevailing wind. Be warned, though that there may be no prevailing wind direction or more than one.

It may be that the wind troubles you only when it blows from a particular direction. In that case you know where to position the shelter. If conditions are windy much of the time and the wind comes from various directions, you may decide to surround the site completely with walls, fences, trees, or hedges.

A solid barrier such as a fence or wall will provide shelter downwind for a distance equal to 10–15 times its height, but as the wind is deflected over the barrier it will form big, strong eddies that may actually increase the wind speed over your plants. A line of trees or a hedge, which allows air to pass through but slows it down, will not stop the wind entirely but it will reduce its speed for a distance downwind equal to 15–20 times its height, and it will not produce eddies in the way a solid barrier does. Remember also, however, that a barrier will also cast shade, so in a small space it may be impossible to shelter the plants from the wind while at the same time allowing them full access to sunshine.

Temporary screens and hurdles

Young, tender plants are especially susceptible to wind damage and may need protection to help them through this stage in their lives. They don't need permanent shelter, but a temporary screen.

This is not a new problem and the traditional solution was to erect one or more hurdles, in North America more often called fence sections or panels. Hurdles are screens, usually about 2 metres wide and between 1–2 metres high. They are woven from willow wands or strips of hazel. These form the weft with poles forming the vertical warp. Willow wands are up to about 1 centimetre in diameter. Hazel strips are wider, up to 2 centimetres. A willow hurdle appears more delicate than a hazel hurdle, but it is less robust. Hurdles are also made from bamboo canes in a wooden frame. Hurdles have been used for many centuries to partition fields and enclose livestock and farm workers used to make them as required. Today you can buy them online. Joined together end to end they make attractive fencing, but deployed individually they will provide very local wind shelter.

As well as being light and easy to move and erect, the point about hurdles is that they are not solid. The wind can blow through them, but it loses energy in doing so, and it does not form eddies on the downwind side.

Hurdles are beautiful and effective, but they aren't cheap and there is an alternative you can make for yourself. You need to erect poles about 9

A traditional hurdle consists of woven sections made from willow or hazel. It makes attractive fencing but is also an effective temporary shelter from the wind.

The wall in the background faces the Sun. It absorbs warmth which it then releases slowly, benefitting the plants growing close to it, while at the same time sheltering them from the wind.

centimetres in diameter spaced at twice the height of your screen—if the screen is to be 2 metres tall the poles should be 4 metres apart. Then fix plastic netting to the poles, making sure it is held taut.

Once the plants are able to withstand the wind the hurdles can be removed and stored.

Walls to store warmth

A solid wall of brick or stone is attractive in its own right, especially when it has aged. But if the wall faces toward the equator it will also provide a warming, welcoming situation for growing plants descended from natives of warmer climes. A wall cossets plants in three ways. In the first place it shelters them against winds blowing from any direction in an arc of about 160 degrees and from eddies in the wind that crosses over the wall, because these reach ground level some distance from the foot of the wall, so do not touch plants growing against or very close to the wall.

The equator-facing wall also reflects sunshine. A wall has an albedo—reflectivity—of about 0.2. This is low, but it means that 20% of the solar radiation striking the wall is reflected. Plants growing against the wall are warmed from one direction by direct sunshine and, to a much lesser but far from insignificant extent, from the opposite direction by reflected sunshine.

HEAT CAPACITY

Substance	Heat capacity (J/g/K)
fresh water	4.19
ice	2.0
asphalt	0.92
brick	0.84
concrete	0.88
granite	0.79
marble	0.88
sand	0.84
wood	1.7

DEWPOINT TEMPERATURE

Dry-Bulb Temperature (°C)	Wet-Bulb Depression (°C)							
	0.5	1.0	1.5	2.0	2.5	3.0	3.5	4.0
−10.0	−12.1	−14.5	−17.5	−21.3	−26.6	−36.3		
−7.5	−9.3	−11.4	−13.8	−16.7	−20.4	−25.5	−34.4	
−5.0	−6.6	−8.4	−10.4	−12.8	−15.6	−19.0	−23.7	−31.3
−2.5	−3.9	−5.5	−7.3	−9.2	−11.4	−14.1	−17.3	−21.5
0.0	−1.3	−2.7	−4.2	−5.9	−7.7	−9.8	−12.3	−15.2
2.5	1.3	0.1	−1.3	−2.7	−4.3	−6.1	−8.0	−10.3
5.0	3.9	2.8	1.6	0.3	−1.1	−2.6	−4.2	−6.1
7.5	6.5	5.5	4.4	3.2	2.0	0.7	−0.8	−2.3
10.0	9.1	8.1	7.1	6.0	4.9	3.	2.5	1.2
12.5	11.6	10.7	9.8	8.8	7.8	6.7	5.6	4.5
15.0	14.2	13.3	12.5	11.6	10.6	9.6	8.6	7.6
17.5	16.7	1.9	15.1	14.3	13.4	12.5	11.5	10.6
20.0	19.3	18.5	17.7	16.9	16.1	15.3	14.4	13.5
22.5	21.8	21.1	20.3	19.6	18.8	18.0	17.2	16.3
25.0	24.3	23.6	22.9	22.2	21.4	20.7	19.9	19.1
27.5	26.8	26.2	25.5	24.8	24.1	23.3	22.6	21.9
30.0	29.4	28.7	28.0	27.4	26.7	26.0	25.3	24.6
32.5	31.9	31.2	30.	2.9	2.3	28.6	27.9	27.2
35.0	34.4	33.8	33.1	32.5	31.9	31.2	30.6	29.9
37.5	36.9	36.3	35.7	35.1	34.4	33.8	33.2	32.5
40.0	39.4	38.8	38.2	37.6	37.0	36.4	35.8	35.1

The wall reflects 20% of the radiation falling upon it and, therefore, it absorbs the other 80%, which it reradiates fairly efficiently. That is because stone and brick have a low heat capacity. The heat capacity of a substance is the amount of energy required to raise its temperature by 1°C (1.8°F), measured in joules per gram per kelvin (1K = 1°C). The table on page 283 gives the heat capacities of a variety of materials.

Its low heat capacity means that the solar energy the wall absorbs quickly begins to raise its temperature. The wall grows warmer and as its temperature rises, it radiates its absorbed energy as long-wave electromagnetic radiation—heat. The radiated heat warms the plants close to the wall and it goes

4.5	5.0	7.5	10.0	12.5	15.0	17.5
−78.6						
−27.7	−41.3					
−18.9	−23.9					
−12.9	−16.1					
−8.1	−10.4	−47.7				
−4.0	−5.8	−21.6				
−0.2	−1.8	−12.8				
3.2	1.9	−6.8	−28.2			
6.5	5.3	−1.9	−14.5			
9.6	8.5	2.3	−7.0	−35.1		
12.6	11.6	6.1	−1.4	−14.9		
15.5	14.6	9.6	3.2	−6.3	−37.5	
18.3	17.5	12.9	7.3	−0.2	−13.7	
21.1	20.3	16.1	11.1	4.7	−4.7	−31.7
23.8	23.1	19.1	14.5	9.0	1.6	−11.1
26.5	25.8	22.1	17.8	12.8	6.6	−2.4
29.2	28.5	24.9	21.0	16.4	11.0	3.9
31.9	31.2	27.7	24.0	19.8	14.9	8.9
34.5	33.9	30.5	26.9	23.0	18.5	13.3

on warming them late into the day, because the wall continues to radiate its warmth as the sun sinks and the air temperature starts to fall.

If your garden has, or could have, a wall facing the sun, you will be able to grow plants against it that would not survive in a more exposed position.

Frost protection

Will there be a frost? The evening weather forecast should warn you, but most radio and TV weather forecasts are very general because they have to cover a large geographic area. So the forecast is only a rough guide and to be certain you'll have to use your own judgement.

TEMPERATURE FOR SPRAYING FOR FROST PROTECTION

Dewpoint temperature (°C)	Wet-Bulb Temperature (°C)					
	−5.0	−4.0	−3.0	−2.0	−1.0	0.0
0.0						0.0
−1.0					−1.0	0.7
−2.0				−2.0	−0.4	1.3
−3.0			−3.0	−1.4	0.2	1.9
−4.0		−4.0	−2.5	−0.9	0.8	2.4
−5.0	−5.0	−3.5	−1.9	−0.4	1.3	2.9
−6.0	−4.5	−3.0	−1.5	0.1	1.8	3.4
−7.0	−4.1	−2.6	−1.0	0.6	2.2	3.9
−8.0	−3.6	−2.1	−0.6	1.0	2.6	4.3
−9.0	−3.3	−1.7	−0.2	1.4	3.0	4.7

Obviously, if the temperature early in the evening is below freezing, then frost, in the sense of freezing temperatures, is already with you. If the temperature is well above freezing, the drop in temperature through the night is unlikely to bring it low enough to cause frost. It is only when the air is already fairly close to freezing that there is a likelihood of frost forming.

Frost forms on nights when the air is moist and still and the sky is clear. If there is a wind, even a light breeze, the risk of frost is reduced, because the constant movement of air replaces cold air close to the ground and plant surfaces with milder air and the mixing will continue for as long as the wind blows.

Nor is frost likely when the sky is overcast. The ground-level temperature falls because surfaces lose heat by radiation, but clouds reflect some of the radiation and absorb and reradiate more, so a cloud cover checks the rate of temperature decrease.

You can estimate how cold the night will be. The temperature falls at a fairly constant rate from sunset until an hour before dawn, when it levels off before starting to rise. Measure and note the air temperature one hour before sunset and again one and two hours after sunset. Plot the three temperatures and times on a graph and extend that rate of fall until one hour before dawn. That final temperature is how cold the night will be.

An alternative rule of thumb states that the dewpoint temperature at sunset is approximately equal to the air temperature the following dawn. To use this you will need to calculate the dewpoint temperature and for that you will need a psychrometer. Basically, this consists of two thermometers one of which has its bulb wrapped in wet muslin. Make sure before you start that the thermometers agree with each other. Then wrap the wet muslin around the wet-bulb thermometer and wave the thermometer about so that air contacts every part of the muslin. Then take a reading from both thermometers. You'll find the wet-bulb thermometer displays a lower temperature than the dry-bulb thermometer. Subtract the wet-bulb temperature from the dry-bulb temperature and the difference is the wet-bulb depression. Now use the table on pages 284–285 to read off the dewpoint temperature. In the column on the left find the dry-bulb temperature, find the wet-bulb depression in the top row, then follow that column down and the dry-bulb temperature row across. Where they intersect is the dewpoint temperature. If that temperature is below freezing, expect frost.

If frost is likely you will need to protect tender plants. If they are in tubs you may be able to bring them indoors. If they are outdoors you can wrap or cover them with fleece. If the plants at risk are too large to protect in this way a sprinkler might protect them. It sounds paradoxical to spray water over plants as a method of frost protection, but it works because as the sprayed water freezes it releases latent heat, ensuring the plant cells remain above freezing temperature even though the plant may be coated in ice. The aim is to hold the plant at a temperature higher than the lowest it will tolerate.

Use the table opposite to determine when to turn on the sprinkler. First calculate the dewpoint temperature. Then find on the table a wet-bulb temperature that is higher than the lowest temperature your plants can tolerate. Follow that column down until it intersects the row from the dewpoint temperature. The temperature at the intersection is the one at which the sprinkler should be turned on.

Mulching

Mulching is a traditional gardening practice. It suppresses weeds and it also moderates extremes of weather. A layer of mulch, ideally about 7 centimetres deep, shades the soil. Sunlight cannot penetrate. That greatly reduces the rate of evaporation from the soil. The water that does evaporate is trapped in the mulch where it condenses and trickles back to the soil. So mulch applied in spring when the soil is still moist is a good protection against summer drought. You shouldn't apply the mulch when the soil is dry, however, because drizzle, light rain, dew, and mist cannot penetrate it. A heavy downpour will deliver water to the underlying soil, but for most of the time the dry soil will remain dry.

Chipped bark makes a good soil mulch, suppressing weeds and protecting these strawberries.

A layer of mulch also provides thermal insulation. Applied in spring it protects the soil against high summer temperatures that might damage roots. Applied in autumn it helps the soil retain for just a little while longer the warmth it absorbed during the summer, because it reduces the loss of heat at night by radiation.

There are two types of mulching material, organic and inorganic, and both conserve moisture and provide thermal insulation. Organic mulches consist of plant material. These decompose, forming part of the soil as they do so. The decomposed mulch contributes to the soil texture and adds some plant nutrients. It must be replenished as it decomposes, of course, but later additions continue to improve the soil. Inorganic mulches do not decompose. They consist of such material as plastic sheeting, landscape fabric, stones, and gravel.

Depending on their colour, inorganic mulches have an additional effect due to their albedo. White or pale stones with a high albedo reflect most of the sunlight falling on them. This insulates the soil very effectively. The material presents an irregular surface, so the reflected light and heat are scattered in all directions. In sunny weather, plants growing above the mulch will be bathed in very intense light, which they will convert to heat as they absorb it, and they will have little or no shade. Such conditions suit some plants but not all.

Black plastic, with a very low albedo, absorbs most of the sunlight falling on it. On a sunny day you can feel how hot it becomes. This warms the soil and at night it takes some time for the temperature of the plastic to fall, so the plastic continues to warm plants growing above it.

Conclusion

OVER THE CENTURIES GARDENERS HAVE LEARNED to deal with whatever the atmosphere throws at them. They enjoy watching their gardens prosper in fine weather and although they may bemoan the losses that result from foul weather, they accept there is little they can do about it. They clear up, salvage what they can, compost the wreckage, and carry on—at times even cheerfully.

The modest purpose of this short book has been to introduce you to what is sometimes a friend and at other times the enemy. Does that help you become a better gardener? I don't know, but I suggest that the old maxim of knowing your enemy applies here as in other areas of life.

Climates are the accumulation of weather and it is the combined effects on the atmosphere and oceans of energy from the Sun and the rotation of the Earth that generate weather. The atmosphere behaves in very complicated ways and much of its behaviour is inherently unpredictable. Those who tell you they can forecast the weather decades from now deceive themselves. The task is impossible. That is less true for climates. Ten years from today I can assure you it will be extremely cold at Vostok Station, Antarctica, but no one can say whether or not there will be a snowstorm there on that day. But even climates change. They have changed throughout the history of our planet and they will continue to change in the years to come. Again, though, and despite the pretended assuredness of some, no one can describe with any confidence the ways they will change. Of one thing we can be sure, however: there will be gardeners raising plants whatever the weather brings.

People seek certainties and often look to science to provide them. But that is not what science does. Science deals with probabilities, not certainties. It seeks explanations for natural phenomena, but all of those explanations are to a greater or lesser extent provisional. That is not a criticism or a drawback. On the contrary, it is the great strength of the scientific endeavour. All knowledge is subject to review, all ideas are open to refutation, and by substituting better and more comprehensive explanations for those found inadequate, so our knowledge grows and advances. There is always more to learn and answers to questions raise more questions. Certainty would be stultifying. Uncertainty is exhilarating.

I hope that my attempts at explanations have entertained you and, perhaps, informed you. And maybe, if I have succeeded beyond what I have any right to expect, I will have aroused in you an enthusiasm to learn more.

Further reading

Allaby, Michael. *Deserts*, revised edition 2008. Facts On File, New York.

———. *Encyclopedia of Weather and Climate*, revised edition 2007. Facts on File, New York.

Allen, R., Tett, S., and Alexander, L. 2008. "Fluctuations in autumn-winter severe storms over the British Isles: 1920 to present." *International Journal of Climatology*, 29, 357–371.

Ashman, M. R., and G. Puri. 2002. *Essential Soil Science*. Blackwell Science, Oxford.

Atmospheric rivers. http://climatesnack.com/2013/06/14/atmospheric-rivers-2/

Atmospheric stability. http://www.aos.wisc.edu/~aalopez/aos101/wk10.html

Australian National Botanic Gardens Centre for Australian National Biodiversity Research. http://www.anbg.gov.au/index.html

Dawson, A., Elliott, L., Noone, S., Hickey, K., Holt, T., Wadhams, P., and Foster, I. 2004. "Historical storminess and climate "see-saws" in the North Atlantic region." *Marine Geology*, 210, 247–259.

Department of Energy and Climate Change. 2013. *Thermal Growing Season in Central England*. https://www.gov.uk/government/uploads/system/uploads/attachment_data/file/192601/thermal_growing_season_summary_report.pdf

Diffusion and osmosis. http://hyperphysics.phy-astr.gsu.edu/hbase/kinetic/diffus.html

Esteves, L. S., Williams, J. J., and Brown, J. M. 2011. "Looking for evidence of climate change impacts in the eastern Irish Sea." *Natural Hazards and Earth System Sciences*, 11, 1641–1656.

Fleming, James Rodger. *Fixing the Sky: The Checkered History of Weather and Climate Control.* 2010. Columbia University Press, New York.

Frost damage. http://www.fao.org/docrep/008/y7223e/y7223e0a.htm

General circulation of the atmosphere. http://cimss.ssec.wisc.edu/wxwise/class/gencirc.html

Germination temperature. http://tomclothier.hort.net/page11.html

Greenhouse effect. http://www.weatherquestions.com/What_is_the_greenhouse_effect.htm

Hamblin, Jacob Darwin. *Arming Mother Nature: The Birth of Catastrophic Environmentalism.* 2013. Oxford University Press, New York.

International Phenological Gardens of Europe. http://www.agrar.hu-berlin.de/fakultaet-en/departments/dntw-en/agrarmet-en/phaenologie/ipg

IPPC. 2012. "Summary for Policymakers." *In Managing the Risks of Extreme Events and Disasters to Advance Climate Change Adaptation. A special report of Working Groups I and II of the Intergovernmental Panel on Climate Change.* Cambridge University Press, Cambridge and New York.

———. *Assessment Report 5: Summary for Policymakers.* http://www.ipcc.ch/report/ar5/wg1/docs/WGIAR5_SPM_brochure_en.pdf

Kew's Millennium Seed Bank. http://www.kew.org/science-conservation/millennium-seed-bank-partnership/about/millennium-seed-bank-faqs

Köppen climate classification. http://xpeditiononline.com/datavis/koppenguide.pdf

Kozlov, M. V., and Berlina, N. G. 2002. "Decline in length of the summer season on the Kola Peninsula, Russia." *Climatic Change*, 54, 387–398.

Lamb, H. H. *Climate History and the Modern World.* 2nd edition 1995. Routledge, London.

———. *Historic Storms of the North Sea, British Isles and Northwest Europe.* 1991. Cambridge University Press, Cambridge.

Linderholm, Hans W. 2006. "Growing season changes in the last century." *Agricultural and Forest Meteorology*, 137, 1–14. http://research.eeescience. utoledo.edu/lees/papers_PDF/Linderholm_2006_AFM.pdf

Little Ice Age. http://academic.mu.edu/meissnerd/4horses.htm

Mountains: globally important ecosystems. http://www.fao.org/docrep/ w9300e/w9300e03.htm

North Atlantic Oscillation. http://ossfoundation.us/projects/environment/ global-warming/north-atlantic-oscillation-nao

Pacific Decadal Oscillation. http://www.drroyspencer.com/global-warming-background-articles/the-pacific-decadal-oscillation/

Photoperiodism. http://tnau.ac.in/eagri/eagri50/PPHY261/lec15.pdf

Plimer, Ian. *Heaven and Earth. Global Warming: The Missing Science.* 2009. Quartet Books, London.

Plutarch. *The Life of Marius, in The Parallel Lives.* http://penelope.uchicago .edu/Thayer/E/Roman/Texts/Plutarch/Lives/Marius*.html

Principles of Frost Protection. http://biomet.ucdavis.edu/frostprotection/ Principles%20of%20Frost%20Protection/FP005.html

Remote sensing. http://www.remss.com/measurements/upper-air-temperature

Rossby waves. http://www.eumetcal.org/euromet/english/nwp/n2d00/ n2d00009.htm

Soil Composition and Formation. http://www.nerrs.noaa.gov/doc/siteprofile/ acebasin/html/envicond/soil/slform.htm

Soil Permeability. ftp://ftp.fao.org/fi/CDrom/FAO_Training/FAO_Training/ General/x6706e/x6706e09.htm

Stefan–Boltzmann law. http://hyperphysics.phy-astr.gsu.edu/hbase/thermo/ stefan.html

Svensmark, Henrik, and Calder, Nigel. 2007. *The Chilling Stars: A New Theory of Climate Change*. Allen & Unwin Pty Ltd, Crows Nest, NSW, Australia.

Taiga or Boreal Forest. http://www.marietta.edu/~biol/biomes/boreal.htm

Teleconnections. http://www.cpc.ncep.noaa.gov/data/teledoc/teleintro.shtml

Thunderstorm Hazards—Hail. http://www.srh.noaa.gov/jetstream/tstorms/hail.htm

Tornado and Storm Research Organisation (TORRO). http://www.torro.org.uk/site/index.php

Tree line. http://www.nature.com/scitable/knowledge/library/global-treeline-position-15897370

US Soil Taxonomy. http://plantandsoil.unl.edu/pages/informationmodule.php?idinformationmodule=1130447032&topicorder=2&maxto=16&minto=1

Welwitschia mirabilis. http://www.plantzafrica.com/plantwxyz/welwitschia.htm

Woodworth, P. L., and Blackman, D. L. 2002. "Changes in extreme high waves at Liverpool since 1768." *International Journal of Climatology*, 22, 697–714.

World Reference Base for Soil Resources. http://www.fao.org/docrep/w8594e/w8594e00.htm

Glossary

ablation The removal of snow and ice by melting, sublimation, or wind.

abscission The action of shedding or detaching, for example of a leaf.

absolute humidity The mass of water vapour present in a specified volume of air, taking no account of variations due to changes in temperature and pressure.

activation energy (energy of activation) The energy a biological system must absorb in order to increase the number of reactive molecules within it and thus initiate a chemical reaction.

adherence The property of water molecules that makes them attach (adhere) to the solid sides of a container.

adiabatic Describes a change in temperature that occurs with no exchange of energy with an outside source.

adret Describes a sloping surface that faces toward the equator.

advection The transport of heat by the movement, usually horizontal, of air or water.

advection fog FOG that forms when warm, moist air moves horizontally over a cool surface.

air frost The condition in which the air temperature is below freezing.

air mass A body of air covering a large area, for example a continent or ocean, within which the physical characteristics of density, temperature, LAPSE RATE, and HUMIDITY are approximately constant at every height.

albedo The proportion of the radiation falling on a surface that is reflected; the reflectivity of a surface.

anabatic wind (upslope wind) A wind that blows uphill.

anion An ion with negative charge.

anticyclone An area of high atmospheric pressure around which air circulates in a clockwise direction in the Northern Hemisphere and anticlockwise in the Southern Hemisphere.

anticyclonic A circular flow of air clockwise in the Northern Hemisphere and anticlockwise in the Southern Hemisphere.

AO *See* Arctic Oscillation.

aphelion The point in its orbit when a planet is farthest from its star. Earth is at aphelion on 4 July.

aquifer A subsurface layer of permeable material that is capable of storing water and through which GROUNDWATER flows. If there is no impermeable material above it, the aquifer is said to be unconfined. If there is an impermeable layer above it, the aquifer is confined.

Arctic Oscillation (AO, Northern Annular Mode, Northern Hemisphere Annular Mode) A periodic change in surface pressure between the arctic and latitudes 37–45° N.

aspect The direction sloping ground faces.

atmospheric boundary layer *See* planetary boundary layer.

atmospheric river A stream of moist air in the upper TROPOSPHERE, up to about 300 kilometres wide and usually 1000–2000 kilometres long, that transports moisture away from the equator.

axial tilt The angle between the Earth's axis of rotation and a line passing through the centre of the Earth at right angles to the PLANE OF THE ECLIPTIC. At present that angle is 23.5 degrees.

Azores high A semi-permanent area of high pressure, a SUBTROPICAL HIGH, that centres over the Azores. Sometimes this area moves to the western side of the North Atlantic, when it is known as the Bermuda high.

backing A change in wind direction in an anticlockwise direction, for example from southerly to easterly.

baumgrenze *See* tree line.

Bermuda high *See* Azores high.

black frost (hard frost) A blackening of plant surfaces that occurs when the temperature falls well below freezing in very dry air. No frost forms on exposed surfaces, but moisture freezes in the plant tissues.

blocking The obstruction of weather systems by a persistent ANTICYCLONE or CYCLONE, causing the systems to pass to the north or south and resulting in a prolonged spell of a particular type of weather.

boundary current An ocean current that flows close to and parallel to a continental coast. Western boundary currents carry warm water and eastern boundary currents carry cool water.

boundary layer A layer of air in which conditions are strongly influenced by contact with a surface.

butterfly effect A metaphor describing an extreme sensitivity to initial conditions, such that a variation too small to detect can cause initially similar atmospheric conditions to develop in entirely different ways.

cap A thin, impermeable surface layer on a soil.

capillarity (capillary attraction) The process by which water moves in any direction through very small spaces.

capillary attraction *See* capillarity.

capillary fringe The layer of partially saturated soil immediately above the WATER TABLE in which water moves upward by CAPILLARITY.

catena (toposequence) A sequence of closely related soils derived from the same PARENT MATERIAL that differ as a result of drainage or some other factor.

cation An ion with positive charge.

cation exchange capacity (CEC) The capacity of a soil for the exchange of CATIONS between positions on SOIL COLLOIDS and between colloids and the soil solution.

CCN *See* cloud condensation nuclei.

CEC *See* cation exchange capacity.

chemical weathering The WEATHERING of rock by chemical reactions between the soil solution and rock minerals.

climate normal A period, commonly of 30 years, over which the daily weather conditions at a specified location are averaged to produce a description of the climate at that place.

clothesline effect The drying effect of wind at the edge of a stand of vegetation.

cloud condensation nuclei (CCN) Minute airborne particles on to which water vapour condenses to form cloud droplets.

coherence The property of water molecules that makes them cling together.

cold front A weather FRONT at which the air behind the front is cooler than the air ahead.

cold lightning Lightning that does not ignite ground fires.

confined aquifer *See* aquifer.

convection cell A vertical circulation in which a fluid is heated from below, rises and cools, and subsides to be heated once more.

CorF *See* Coriolis effect.

coriaceous Leathery.

Coriolis effect (CorF) The deflection that a moving body experiences as it crosses the Earth's surface, to the right in the Northern Hemisphere and to the left in the Southern Hemisphere. The magnitude of the effect is zero at the equator and reaches a maximum at the poles.

cotyledon (seed leaf) The embryonic leaf in a plant seed. There is one cotyledon in monocots, two in dicots, and two or more in gymnosperms.

cuticle The thin, waxy, outer skin of the leaves and/or stems of certain plants.

cyclone 1. A region of low atmospheric pressure around which air circulates anticlockwise in the Northern Hemisphere and clockwise in the Southern Hemisphere. 2. A TROPICAL cyclone that forms in the Bay of Bengal. *See also* extratropical cyclone.

cytoplasmic matrix *See* cytosol.

cytosol (intracellular fluid, cytoplasmic matrix) The liquid that fills cells.

DALR *See* dry adiabatic lapse rate.

dangerous semicircle The side of a TROPICAL CYCLONE farthest from the equator, where the winds are strongest.

dart leader The first stage of the second and all subsequent strokes that comprise a lightning stroke.

denaturation A change in the structure of a protein or nucleic acid that alters its biological activity.

deposition The formation of ice directly from water vapour.

derecho A storm with winds of up to 130 km/h that blow in a straight line.

dewpoint The temperature at which air would become saturated if it were cooled with no change in its moisture content or in atmospheric pressure.

dewpoint lapse rate The rate at which the DEWPOINT temperature decreases with increasing height due to the fall in atmospheric pressure. It is about 2°C (3.6°F) per kilometre.

drainage wind *See* katabatic wind.

dry adiabatic lapse rate (DALR) The rate at which rising unsaturated air cools ADIABATICALLY with height, and subsiding air warms. It is 9.8°C (17.6°F) per kilometre.

eccentricity The extent to which a planetary orbit deviates from a circle.

ECS *See* equilibrium climate sensitivity.

El Niño A change in the prevailing trade winds in the equatorial South Pacific, and the wind-driven surface ocean current that ordinarily flows westward. The winds weaken or reverse direction and warm water flows eastward, accumulating off the South American coast.

ELR *See* environmental lapse rate.

eluviation The process in which material is removed in suspension from upper SOIL HORIZONS and partially deposited in lower horizons.

endosperm In the seed of an angiosperm, a store of nutrients to sustain the seedling.

energy of activation *See* activation energy.

ENSO A complete cycle of EL NIÑO and LA NIÑA with the associated SOUTHERN OSCILLATION; an El Niño–southern oscillation cycle.

entrainment Mixing between a body of air and the air surrounding it.

environmental lapse rate (ELR) The rate at which temperature decreases with height as measured at a particular place and time.

enzyme A chemical compound, protein or mainly protein, that is produced by a living cell and that accelerates a biochemical reaction while remaining unaltered. The enzyme either lowers the ACTIVATION ENERGY or facilitates an alternative chemical pathway that has a lower activation energy.

epidermis The outermost layer of tissue, in plants one cell thick.

equilibrium climate sensitivity (ECS) The response of the global mean surface air temperature to a doubling of the atmospheric concentration of carbon dioxide.

equinox One of the two dates each year when the Sun is directly overhead at noon over the equator, and day and night are of equal length throughout the world. At present the equinoxes fall on 21 March and 21 September.

erg A large expanse of sand dunes in the Sahara Desert, a sand sea.

evapotranspiration Evaporation and TRANSPIRATION considered together.

exfoliation (onion-skin weathering) The peeling away of the outer layers of rock due to thermal stress when sunshine causes the surface to expand more than the underlying rock.

extratropical cyclone (depression) A CYCLONE that forms in a frontal system outside the tropics. *Compare* tropical cyclone.

eye The calm centre of a cyclonic storm, especially a TROPICAL CYCLONE.

eyewall The towering CUMULONIMBUS storm clouds that surround the EYE of a TROPICAL CYCLONE, delivering the heaviest precipitation and strongest winds.

fallstreaks *See* virga.

fall wind *See* katabatic wind.

fern frost Frost that forms patterns reminiscent of fern fronds, usually on windowpanes. It is caused by the freezing of water that condensed onto the glass.

Ferrel cell Part of the circulation of the atmosphere, comprising a midlatitude cell driven by the HADLEY and POLAR CELLS, in which air rises at the POLAR FRONT, moves toward the equator, and subsides in the subtropics.

fetch The distance air moves across a land or water surface.

fog Precipitation in the form of cloud that extends to the surface and reduces visibility to less than 1 kilometre.

fog drip Water that drips to the ground from FOG that has wetted tree branches and other structures.

föhn wind A warm, dry wind that blows on the lee side of a mountain, most often in spring.

freezing fog FOG that forms when the air temperature, and surfaces exposed to the air, is below freezing. Fog droplets freeze instantly on contact with the cold surfaces.

freezing nuclei Particles on which water vapour will form ice crystals by DEPOSITION.

front A boundary between two AIR MASSES.

frost hazard The likelihood that plants will be damaged by frost.

frost point The temperature at which water vapour turns directly to ice.

frozen fog FOG that forms when supercooled water droplets freeze to form ice crystals that grow at the expense of remaining unfrozen droplets.

funnelling The channelling of wind along a steep-sided valley or city street lined by tall buildings. This accelerates the wind and reduces the air pressure slightly.

geophyte A plant that survives harsh conditions by means of underground organs that store food and produce buds from which new shoots arise when conditions above ground are favourable.

geostrophic wind A wind that blows parallel to the ISOBARS. It occurs at a height where it is unaffected by friction with the surface.

gleying (gleyzation) The soil process, resulting from prolonged water-logging, in which iron compounds are reduced and move downward, producing grey horizons with red mottling.

gleyzation *See* gleying.

graupel (soft hail) Precipitation in the form of opaque, white, ice pellets, most 2–5 millimetres in size, that are soft enough to smash into fragments or flatten when they strike a hard surface.

gravity wind *See* katabatic wind.

ground frost The condition in which the ground temperature is below freezing, but the air temperature is above freezing.

groundwater Water held in the saturated zone of the soil, above a layer of impermeable material.

gust A sudden, sharp, short-lived increase in the wind speed.

gust front A line ahead of an advancing storm where warm, moist air is being drawn into the base of the cloud, producing strong wind GUSTS.

gustnado A small TORNADO that develops in a GUST FRONT.

gyre An approximately circular flow of currents in each of the ocean basins, moving clockwise in the Northern Hemisphere and anticlockwise in the Southern Hemisphere.

Hadley cell Part of the circulation of the atmosphere, comprising a CONVECTION CELL in which warm air rises over equatorial regions, moves away from the equator, and subsides over the subtropics.

hailshaft A column of falling hailstones visible below a cloud.

hailstreak A strip of ground that is completely covered by fallen hailstones.

hailswath An area of ground that is partly covered by hailstones.

hammada A desert surface comprising exposed bedrock and large boulders.

hard frost *See* black frost.

hardiness The extent to which a plant withstands low temperatures.

heat capacity (thermal capacity) The amount of energy a substance must absorb in order for its temperature to rise by a specified amount, usually 1°C (1.8°F). Heat capacity may refer to a unit mass of the substance, when it is known as the specific heat capacity (c) or a unit amount of the substance, when it is known as the molar heat capacity (C_m).

hill fog (upslope fog) FOG encountered on hillsides that forms in rising air cooled to below its DEWPOINT temperature.

horse latitudes Regions at about 30° latitude in both hemispheres where subsiding air in the SUBTROPICAL HIGHS produces light and variable winds.

hot lightning Lightning that ignites ground fires.

humidity The amount of water vapour present in the air. *See* absolute humidity, mixing ratio, relative humidity, specific humidity.

hurricane *See* tropical cyclone.

hydraulic head The height of a water surface above a specified datum level.

hyperosmotic *See* hypertonic.

hypertonic (hyperosmotic) Describes a cell in which the OSMOTIC PRESSURE is higher than it is outside the cell.

hyposmotic *See* hypotonic.

hypotonic (hyposmotic) Describes a cell in which the OSMOTIC PRESSURE is lower than it is outside the cell.

hypodermis A layer of cells, often strengthened, immediately below the EPIDERMIS.

Iceland low A semi-permanent area of low pressure centred over Iceland.

illuviation The process in which materials are washed into a soil or soil layer.

imbibition The absorption (imbibing) of water.

index cycle A sequence of wave development in the polar front JET STREAM that ends with the formation of isolated ANTICYCLONES and CYCLONES.

infiltration capacity The maximum rate at which soil or other material can absorb rainfall.

intertropical convergence zone (ITCZ) A belt close to the equator and surrounding the Earth where the trade winds of the North and South Hemispheres meet and their convergence causes air to rise.

intracellular fluid *See* cytosol.

inversion (temperature inversion) A layer in the atmosphere within which the temperature remains constant or increases with increasing altitude.

isobar A line on a map joining places of equal atmospheric pressure.

isobaric slope *See* pressure gradient.

isobaric surface A surface where the atmospheric pressure is constant throughout.

isotonic Describes a cell in which the OSMOTIC PRESSURE is equal to that outside the cell.

ITCZ *See* intertropical convergence zone.

jet stream A long, winding ribbon of wind blowing at an average 100 km/h but sometimes at up to 400 km/h, that is located close to the TROPO-PAUSE. There are several jet streams. Typically they are thousands of kilometres long, hundreds of kilometres wide, and often no more than 5 kilometres deep.

katabatic wind (drainage wind, fall wind, gravity wind) A cool wind that blows downhill, consisting of chilled air that is sinking beneath less dense air below.

killing frost A fall in temperature that kills plants or prevents them from reproducing.

kinetic energy The energy of motion, equal to $\frac{1}{2}\,mv^2$, where m is the mass of the body and v its velocity. If the body is rotating its kinetic energy is equal to $\frac{1}{2}\,I\Omega^2$, where I is the moment of inertia and Ω is the angular velocity.

krummholz Vegetation comprising stunted, gnarled trees and shrubs growing close to the TREE LINE.

landspout (land waterspout) A small TORNADO that develops from a non-SUPERCELL storm cloud.

land waterspout *See* landspout.

La Niña The opposite of an EL NIÑO event, in which the trade winds over the South Pacific intensify and the WARM POOL in the west deepens.

lapse rate The rate at which air temperature decreases with increasing altitude.

latent heat Heat that is absorbed and released when a substance changes phase between solid and liquid, liquid and gas, or solid and gas.

leaching The removal from soil of substances in solution.

lenticel A pore in the stem of a woody plant that allows air to reach tissues below the surface and through which moisture is lost by TRANSPIRATION.

lessivage The downward movement of soil mineral particles, especially clay.

lifting condensation level The height at which rising air cools to the DEWPOINT temperature and water vapour starts to condense.

lightning channel The path, about 20 centimetres wide, followed by a lightning stroke.

lower atmosphere The TROPOSPHERE.

mass mixing ratio *See* mixing ratio.

mb *See* millibar.

mechanical weathering The WEATHERING of rock by the physical action of wind, rain, and changes of temperature.

meridional flow A flow of air that is approximately parallel to the lines of longitude (meridians).

mesocyclone A rotating mass of air at the centre of a SUPERCELL storm cloud.

mesopause The boundary between the STRATOSPHERE and MESOSPHERE, its base about 80 kilometres above sea level.

mesosphere The layer of the atmosphere between the STRATOPAUSE and MESOPAUSE.

microclimate The climate of a restricted area, which may be markedly different from that of the surrounding region.

millibar (mb) A unit of measurement of pressure often used on weather maps. It is equal to 0.001 bar and 100 pascals.

mist Cloud at surface level in which horizontal visibility, although reduced, is more than 1 kilometre.

mixing ratio (mass mixing ratio) The ratio of the mass of any gas present in the air to a unit mass of dry air without that gas. It is most often used to report HUMIDITY as grams H_2O per kilogram of air.

molar heat capacity *See* heat capacity.

NAO *See* North Atlantic Oscillation.

North Atlantic Oscillation (NAO) A periodic change in the distribution of surface pressure between the ICELAND LOW and AZORES HIGH.

Northern Annular Mode *See* Arctic Oscillation.

Northern Hemisphere Annular Mode *See* Arctic Oscillation.

onion-skin weathering *See* exfoliation.

orographic Pertaining to mountains or high ground.

orthodox seed A seed that can be stored for a long period under controlled conditions.

osmosis The process in which solvent crosses a semipermeable membrane from a region of low SOLUTE concentration to one of higher solute concentration until the two SOLUTIONS are of equal concentration.

osmotic pressure The pressure that is required to prevent solvent molecules crossing a semipermeable membrane from a region of low solute concentration to one of higher concentration.

overland flow *See* surface runoff.

parenchyma Plant tissue consisting of unspecialized cells with air spaces between them.

parent material The rock from which the mineral component of a soil is derived.

PE *See* potential evapotranspiration.

pericarp The wall of a fruit.

perihelion The point in its orbit when a planet is closest to its star. Earth is at perihelion on 4 January.

petiole The stalk that attaches a leaf to the plant stem.

PGF *See* pressure-gradient force.

phenology The study of timing of natural events, for example leaf opening, flowering, setting of fruit, and so on.

photoperiodism The physiological response of an organism, especially a plant, to changes in the relative lengths of day and night, determining, for example, the time of flowering and leaf ABSCISSION.

plane of the ecliptic An imaginary disc with a circumference defined by the path of the Earth's orbit about the Sun.

planetary boundary layer (atmospheric boundary layer, surface boundary layer) The BOUNDARY LAYER in which conditions and the movement of air are strongly influenced by the proximity of the land or sea surface. The depth of the layer varies but seldom exceeds about 500 metres.

polar cell Part of the circulation of the atmosphere, comprising a cell in which cold air subsides over the pole, moves away from the pole at low level, rises at the POLAR FRONT, and returns to the pole.

polar front The midlatitude boundary between tropical and polar air.

potential evapotranspiration (PE) The amount of water that would evaporate from the ground and be removed by TRANSPIRATION if the supply of water were unlimited.

precession of the equinoxes A change in the dates of APHELION and PERIHELION over a cycle of 22,000 to 26,000 years due to the change in the Earth's AXIAL TILT, affecting the position in the orbit at which the EQUINOXES and SOLSTICES occur.

pressure gradient (isobaric slope) A change in atmospheric pressure over a horizontal distance.

pressure-gradient force (PGF) The acceleration force acting on air located on the high-pressure side of a PRESSURE GRADIENT.

psychrometer An instrument that is used to determine the DEWPOINT temperature and RELATIVE HUMIDITY.

radiation fog FOG that forms on clear nights when the air is moist. The ground radiates the warmth it accumulated by day and its temperature falls sharply, chilling the air in contact with it to below the DEWPOINT temperature.

radicle The embryonic root contained in a seed.

rain shadow The land on the lee side of a mountain or mountain range. Moist air is forced to rise on the windward side. The air cools, and its water vapour condenses to form cloud and precipitation. As the air subsides on the lee side it warms by compression and its RELATIVE HUMIDITY decreases further.

rainshaft The narrow belt of heavy precipitation at the rear of a storm cloud.

recalcitrant seed A seed that cannot tolerate desiccation and, therefore, cannot be stored for a long period.

reg A desert surface comprising small pebbles and gravel.

regolith A surface layer of mineral particles.

relative humidity (RH) The ratio of the mass of water vapour present in a unit mass of dry air to the mass required to saturate that air, expressed as a percentage.

return stroke A very luminous lightning stroke triggered by a STEPPED LEADER or DART LEADER.

RH *See* relative humidity.

ridge A tongue-like extension of high pressure.

rime frost *See* rime ice.

rime ice (rime frost) A layer of frost that is white with an irregular surface. It forms when supercooled droplets freeze on to a surface that is at or below freezing temperature, or by DEPOSITION.

safe semicircle The side of a TRIOPICAL CYCLONE closest to the equator, where the winds are lightest.

SALR *See* saturated adiabatic lapse rate.

saturated adiabatic lapse rate (SALR) The rate at which rising saturated air cools ADIABATICALLY and subsiding air warms. This varies according to air temperature from about 5–9°C (9–16°F) per kilometre, with an average value of about 6°C/km (11°F).

saturation The condition in which the moisture content of the air is at a maximum—the RELATIVE HUMIDITY is 100%.

sclerophyllous Describes leaves that are small, thick, and leathery.

seed coat *See* testa.

seed leaf *See* cotyledon.

serotiny The retention by some tree species of seeds in pods or cones, sometimes for several years, until a major event, usually the heat from a fire, causes them to be released. The seeds germinate in soil enriched by ash and in the absence of competition.

SMT *See* soil moisture tension.

soft hail *See* graupel.

soil colloid A mineral or humus particle that is microscopic in size but larger than a molecule, with a very large surface area in relation to its volume that carries a permanent or variable electrostatic charge.

soil horizon A soil layer in which the composition and characteristics are fairly well defined.

soil moisture tension (SMT) The force that causes water to rise through small spaces.

soil profile A vertical section through a soil from the surface to the bedrock.

solstice One of the two dates each year when the Sun is directly overhead at noon over one or other of the tropics, and in each hemisphere the difference between hours of daylight and darkness reach their most extreme. At present the solstices fall on 21 June and 21 December.

solute A substance that is dissolved in another substance, the SOLVENT.

solution An homogenous mixture of two substances in which the molecules of one, the SOLUTE, are dispersed evenly throughout the other, the SOLVENT.

solvent A substance in which another substance, the SOLUTE, is dispersed to form an homogenous mixture, a SOLUTION.

southern oscillation A periodic change in the distribution of surface air pressure over the South Pacific that is linked to EL NIÑO events.

specific heat capacity *See* heat capacity.

specific humidity The ratio of the mass of water vapour present in the air to a unit mass of that air including the water vapour.

spray ring A cloud of spray at the base of a waterspout.

squall A storm that generates brief episodes of increased wind speed.

squall line A series of SQUALLS that forms as a line along a COLD FRONT but then detaches from the front and moves ahead of it.

steam fog FOG that forms when cold air crosses the surface of water that is warmer.

stepped leader The first, barely visible, stage in a lightning stroke.

stoma *See* stomata.

stomata (sing. stoma) Pores in plant leaves through which gases are exchanged and moisture is lost by TRANSPIRATION.

stratopause The boundary between the STRATOSPHERE and MESOPHERE. Its height varies from about 50–60 kilometres with season and latitude.

stratosphere The layer of the atmosphere that lies between the TROPOPAUSE and STRATOPAUSE.

sublimation The direct change from ice to water vapour without passing through the liquid phase.

subtropical high One of several regions of high surface pressure located over oceans in the subtropics. They are most strongly developed in summer and affect climates by blocking depressions travelling eastward in middle latitudes.

suction vortex A small tornado that develops around the edge of a large tornado, caused by turbulence in the airflow into the main tornado.

supercell A storm cloud that contains a single CONVECTION CELL that may develop into a MESOCYCLONE.

supercooled The condition of water droplets that remain liquid at temperatures below 0°C (32°F).

supersaturation The condition of air in which the RELATIVE HUMIDITY exceeds 100%.

surface boundary layer *See* planetary boundary layer.

surface runoff (overland flow) Rainwater that accumulates on or flows downhill across the land surface when the rate at which rain is falling exceeds the INFILTRATION CAPACITY of the soil.

teleconnection A recurring variation in weather pattern that affects a very large geographical area, for example an entire ocean basin or continent, and that persists for weeks or months, or sometimes several consecutive years, so that weather conditions are linked in places separated by vast distances.

temperature inversion *See* inversion.

testa (seed coat) The outer covering of a seed, comprising several layers.

thermal capacity *See* heat capacity.

thermal wind A wind generated when there is a large change in temperature over a short horizontal distance. The JET STREAMS are thermal winds.

thermopause The upper boundary of the THERMOSPHERE at a height ranging from 500–1000 kilometres depending on the intensity of sunlight.

thermosphere The uppermost layer of the atmosphere, above the MESOPAUSE.

thigmomorphogenesis The morphological response of plants to physical contact, for example to wind.

toposequence *See* catena.

tornado A rotating vortex of air that forms inside a CUMULONIMBUS cloud, usually from a MESOCYCLONE, and that extends downward through the cloud base, becoming a tornado when it touches the ground.

transpiration The evaporation of water from plants through STOMATA and LENTICELS.

tree line (baumgrenze) The upper altitudinal or latitudinal boundary beyond which the climate is too cold for tree growth.

tropical cyclone A CYCLONE that forms in the tropics by strong convection and not involving weather FRONTS. It generates heavy rain and high winds and is known as a hurricane if it develops in the North Atlantic or Caribbean, a typhoon in the Pacific, and a cyclone in the Bay of Bengal.

tropopause The boundary between the TROPOSPHERE and STRATOSPHERE, at about 16 kilometres at the equator, 11 kilometres in middle latitudes, and 8 kilometres at the poles.

troposphere The lowest layer of the atmosphere in which air is thoroughly mixed. It extends from the surface to the TROPOPAUSE.

trough A tongue-like extension of low pressure.

typhoon *See* tropical cyclone.

ubac Describes a sloping surface that faces away from the equator.

unconfined aquifer *See* aquifer.

Universal Time (UT) The name adopted in 1928 by the International Astronomical Union to replace Greenwich Mean Time. UT is counted from midnight.

upper atmosphere All of the atmosphere above the TROPOPAUSE.

upslope fog *See* hill fog.

upslope wind *See* anabatic wind.

urban canopy The rooftops in an urban area beneath which atmospheric conditions are strongly influenced by the many MICROCLIMATES in and around parks, gardens, streets, and buildings.

urban dome A dome-shaped CONVECTION CELL that forms beneath a temperature INVERSION over an urban area, most commonly in winter.

valley fog RADIATION FOG that forms in valleys when air chilled by contact with the valley sides sinks by gravity.

veering A change in wind direction in a clockwise direction, for example from southerly to easterly.

vernalization The practice of chilling seeds prior to sowing so they will germinate in autumn, remain dormant through the winter, and resume growth in early spring.

virga (fallstreaks) A semi-transparent veil beneath the base of a cloud that extends partway to the surface. It consists of precipitation falling into unsaturated air and evaporating.

vorticity The tendency of a fluid moving across the surface of the Earth to follow a circular path around an axis at right angles to its direction of rotation.

wall cloud A protrusion of a MESOCYCLONE beneath the base of a CUMULONIMBUS cloud. It turns slowly and indicates a strong likelihood of an imminent TORNADO.

warm conveyor belt A stream of air that forms in the air rising up an advancing warm front that transports the moisture ahead of the front.

warm front A weather FRONT at which the air behind the front is warmer than the air ahead.

warm pool A body of warm water in the region of Indonesia consisting of water driven westward by the South Equatorial Current.

warm sector The warm air between a WARM FRONT and COLD FRONT.

water table The upper surface of GROUNDWATER.

weathering The processes of CHEMICAL WEATHERING and MECHANICAL WEATHERING that reduce rock eventually to small grains.

white dew Frost formed when dewdrops freeze.

wind shear A change in wind speed and/or direction with vertical or horizontal distance.

xylem Plant tissue through which water is transported from the soil to all parts of a plant.

zonal flow A flow of air that is approximately parallel to lines of latitude.

zonal index A number calculated from the difference in atmospheric pressure between latitudes 33° N and 55° N and used to measure the strength of midlatitude westerly winds and the alignment of the polar front JET STREAM.

Photo and illustration credits

Photo Credits

Pages 2, Wikimedia/Thomas Bresson; 7, Andrzej Kubik/Shutterstock.com; 10, Wikimedia/KsØstm; 12, Wikimedia/Paxson Woelber; 15 (left), Plamen/ Shutterstock.com; (right), Denis Burdin/Shutterstock.com; 22, Semork/Shutterstock.com; 27, Allen McDavid Stoddard/Shutterstock.com; 34, iStock.com/ simonkr; 40, Wikimedia/Ks0stm; 41, Flickr/Suresh Krishna; 43, Wikimedia/ Ton Rulkens; 44, Galyna Andrushko/Shutterstock.com; 45, Wikimedia/Amble; 51, Alexandre Rotenberg/Shutterstock.com; 59, Flickr/edward stojakovic; 67, Mikhail Iakovlev/123rf; 70, donyanedomam/123rf; 72, Helen Hotson/Shutterstock.com; 75, Wikimedia/Anthonares; 83, Flickr/David Trood; 84, Wikimedia/ Henry Thew Stephenson; 95, NASA/SDO; 97, Flickr/Ashley Perkins; 100, Wikimedia/Christian Jansky; 102 (left), IWM/Royal Air Force official photographer - Goochild A (Flt Lt); (right), Flickr/Frederic Henri; 103, Wikimedia/ANKAWÜ; 104, Albert Russ/Shutterstock.com; 116 (top left), Wikimedia/Rasbak; (top right), Wikimedia/LivingShadow; (middle left), Wikimedia/Famartin; (middle right), Wikimedia/Thomas Bresson; (bottom left), Wikimedia/Miguel Andrade; (bottom right), Sergey Krasnoshchokov/Shutterstock.com; 122, Vladimir Salman/123rf; 124, Wikimedia/Brocken Inaglory; 125, Maxim van Asseldonk/Shutterstock.com; 127, Wikimedia/Ecomuseo di Parabiago; 128 (left), Robert Zp/ Shutterstock.com; (right), Wikimedia/Joe Calzarette; 129, Wikimedia/MSha; 131 (left), MarcelClemens/Shutterstock.com; (right), Wikimedia/Dysprosia; 135, Flickr/Mary Sullivan; 138, Flickr/Corey Templeton; 140, Albert Russ/Shutterstock.com; 142, M. Pellinni/Shutterstock.com; 146, donyanedomam/123rf; 151 (left), Wikimedia/Justin Hobson; (right), Totajla/Shutterstock.com; 154,

Wikimedia/Michael Mancino/FEMA; 157, Wikimedia/Jeff Schmaltz, MODIS Rapid Response Team, NASA/GSFC; 158, Jenny Lilly/Shutterstock.com; 162, Wikimedia/Stavros1; 164, Anna Eshelman; 167 (left), Fremme/Shutterstock.com; (right), steve estvanik/Shutterstock.com; 168, Florin Stana/Shutterstock.com; 170, Eddie Toro/123rf; 172, Jennifer Griner/Shutterstock.com; 182, Flickr/catpicturelady; 185, American Spirit/Shutterstock.com; 188, The Old Curiosity Shop, Portsmouth Street, Kingsway, London (b/w photo), English Photographer (19th century) / Sean Sexton Collection / Bridgeman Images; 190, Jenny Lilly/Shutterstock.com; 193, diixiib/Shutterstock.com; 203, Flickr/Gopal Venkatesan; 208, Wikimedia/USDA Natural Resources Conservation Service; 210, Wikimedia/Brocken Inaglory; 224, Chantal Ringuette/123rf; 226, Wikimedia/Mitch Barrie; 227, Mark Baldwin/Shutterstock.com; 230, llaszlo/Shutterstock.com; 231, Jacek Nowak/123rf; 234 (left), Dario Lo Presti/Shutterstock.com; (right), Wikimedia/Bernard Gagnon; 235, Wikimedia/Forest & Kim Starr; 237, Anton Foltin/Shutterstock.com; 238, Flickr/Global Crop Diversity Trust; 242 (left), eelnosiva/Shutterstock.com; (right), Wikimedia/Forest & Kim Starr; 244, Wikimedia/Eric Bajart; 249, Wikimedia/Matt Lavin; 251, Flickr/Peyri Herrera; 252, Wikimedia/Thomas Bresson; 255, Patrik Mezirka/Shutterstock.com; 258, Wikimedia/Bob Beale; 259, Flickr/Rexness; 261, Wikimedia/Calimo; 262, Wikimedia/Adbar; 263 (top left), Wikimedia/H. Zell; (top right), Wikimedia/Casliber; (bottom), Wikimedia/Abu Shawka; 265, Wikimedia/Alosh Bennett; 267, Darryl Brooks/123rf; 268, Kiwisoul/Shutterstock.com; 271, Flickr/Miguel Vieira; 272 (left), PavelSvoboda/Shutterstock.com; (right), nutsiam/Shutterstock.com; 273, Pi-Lens/Shutterstock.com; 275, BMJ/Shutterstock.com; 276, Mythos/iStock.com; 279, Marek Uliasz/123rf; 280, Cloudia Spinner/Shutterstock.com; 283 (left), Anest/Shutterstock.com; (right), Andrey tiyk/Shutterstock.com; and 288, sanddebeautheil/Shutterstock.com

Illustration Credits

Alan Bryan: page 229
Dave Carlson: pages 216, 221, 228, 250
Anna Eshelman: pages 76, 81, 86, 94, 111, 136, 137, 197, 213, 241, 243, 245, 246
Kate Francis: pages 194, 196, 205, 223, 247, 256
Mike Morganfeld: pages 13, 14, 17, 18, 19, 23, 24, 26, 28, 32, 33, 35, 38, 47, 56, 62, 78, 107, 108, 109, 148, 161, 233, 257, 260, 269, 270
Arthur Mount: pages 42, 53, 57, 58, 61, 113, 114, 121, 123, 126, 132, 141, 144, 145, 147, 152, 153, 154, 155, 159, 160, 166, 171, 175, 178, 179, 181, 184, 187, 200, 201, 202, 204, 249, 264, 266

Conversion tables

Metres	Feet
0.3	1
0.6	2
0.9	3
1.2	4
1.5	5
1.8	6
2.1	7
2.4	8
2.7	9
3	10
6	20
9	30
12	40
15	50
18	60
21	70
24	80
27	90
30	100

Centimetres	Inches
2.5	1
5	2
8	3
10	4
13	5
15	6
18	7
20	8
23	9
25	10
51	20
76	30
100	40
130	50
150	60
180	70
200	80
230	90
250	100

Temperatures

degrees Celsius = $\frac{5}{9} \times$ (degrees Fahrenheit − 32)

degrees Fahrenheit = ($\frac{9}{5} \times$ degrees Celsius) + 32

To convert length:	Multiply by:
Metres to yards	1.09
Centimetres to inches	0.39
Millimetres to inches	0.039
Centimetres to feet	0.033

Index

savannah, 68, 268, 272

Schaefer, Vincent, 101

Schedonnardus paniculatus, 256

Schnidejoch Pass, 72

sclerophyllous plants, 239–240

Scotland, 84, 88, 120

Scots pine, 274

screens, 282–283

scrub oak, 263

sea breezes, 122, 124

sea fogs, 124

Sea People, 82

seasons, 16

sea surface temperatures, 36–37

seaweed, 27

sedges, 12

Sedum sp., 243

seeds, 220–223, 237–239

seeps, 203

Sefton, England, 89

Sempervivum sp., 243

sheet lightning, 141

Sheringham, Norfolk, 162

short-day plants, 247–248

sicklepod, 236

Silene stenophylla, 237

silktassel, 263

silt, 196, 206

silver pampas grass, 271

Simpson Desert, 38

sleet, 115, 136

slopes, 167–169, 174, 176

smog, 185, 188

smooth cordgrass, 248

smudge pots, 102

snow, 19–20, 25, 35, 44–45, 64, 115,
133–139

snowdrifts, 180–183

snow eater, 177

snow fences, 182

snowflakes, 136–138

snowline, 176–178

snow to water ratio, 182–183

soil classifications, 197–199

soil condition, 208

soil formation and aging, 192–195

soil permeability classes, 207

Soil Survey for England and Wales, 198

soil temperature, 220–223

soil textural triangle, 197

soil types, 196–199

solar radiation, 15, 17–21, 49–50,
167–169, 173, 283–284

solstices, 15, 17, 18

Somerset Levels, 96

Sorbus sp., 274

Sorghastrum nutans, 271

South Africa, 261, 268, 271–272

South America, 31, 32, 74, 78, 82, 173,
268, 271, 272, 273, 274

South Atlantic Current, 26

South Equatorial Current, 26–27, 31

Southern Hemisphere, 59, 66, 108,
169, 178

southern oscillation, 33, 94

South Pacific Current, 27

South Pole, 46

Spartina alterniflora, 248, 250

spatial dendrites, 136

spinach, 248

spinning cups anemometer, 169, 170

Spörer, Gustav, 95

Spörer Minimum, 95

spreading hogweed, 236

spring, 91

spring flowers, 230–232, 275

About the author

Michael Allaby is an enthusiastic, prolific, award-winning science writer who has written, edited, or co-authored over 100 books on environmental science. Of these, 17 were about atmospheric science. His 2-volume *Encyclopedia of Weather and Climate* and *Dangerous Weather: Hurricanes* won awards and *DK Guide to Weather* won the 2001 Junior Prize of the Aventis Prizes for Science Books. His *Plants and Plant Life* won *Booklist* Editor's Choice for 2001. He is editor of four science dictionaries for Oxford University Press. Before becoming a full-time writer in 1973 he worked in the police force, the RAF, and as an actor. See his website www. michaelallaby.com for more information.

© ANDREW GRAHAM-WEALL